Learn PySpark

Build Python-based Machine Learning and Deep Learning Models

Pramod Singh

Apress®

Learn PySpark: Build Python-based Machine Learning and Deep Learning Models

Pramod Singh
Bangalore, Karnataka, India

ISBN-13 (pbk): 978-1-4842-4960-4   ISBN-13 (electronic): 978-1-4842-4961-1
https://doi.org/10.1007/978-1-4842-4961-1

Copyright © 2019 by Pramod Singh

Managing Director, Apress Media LLC: Welmoed Spahr
Acquisitions Editor: Celestin Suresh John
Development Editor: James Markham
Coordinating Editor: Aditee Mirashi

Cover designed by eStudioCalamar

Cover image designed by Freepik (www.freepik.com)

Distributed to the book trade worldwide by Springer Science+Business Media New York, 233 Spring Street, 6th Floor, New York, NY 10013. Phone 1-800-SPRINGER, fax (201) 348-4505, e-mail orders-ny@springer-sbm.com, or visit www.springeronline.com. Apress Media, LLC is a California LLC and the sole member (owner) is Springer Science+Business Media Finance Inc (SSBM Finance Inc). SSBM Finance Inc is a **Delaware** corporation.

For information on translations, please e-mail rights@apress.com, or visit www.apress.com/rights-permissions.

Apress titles may be purchased in bulk for academic, corporate, or promotional use. eBook versions and licenses are also available for most titles. For more information, reference our Print and eBook Bulk Sales web page at www.apress.com/bulk-sales.

Any source code or other supplementary material referenced by the author in this book is available to readers on GitHub via the book's product page, located at www.apress.com/978-1-4842-4960-4. For more detailed information, please visit www.apress.com/source-code.

Printed on acid-free paper

I dedicate this book to my wife, Neha, my son, Ziaan, and my parents. Without you, this book wouldn't have been possible. You complete my world and are the source of my strength.

Table of Contents

About the Author

 **Pramod Singh** has more than 11 years of hands-on experience in data engineering and sciences and is currently a manager (data science) at Publicis Sapient in India, where he drives strategic initiatives that deal with machine learning and artificial intelligence (AI). Pramod has worked with multiple clients, in areas such as retail, telecom, and automobile and consumer goods, and is the author of *Machine Learning with PySpark*. He also speaks at major forums, such as Strata Data, and at AI conferences.

Pramod received a bachelor's degree in electrical and electronics engineering from Mumbai University and an MBA (operations and finance) from Symbiosis International University, in addition to data analytics certification from IIM–Calcutta.

Pramod lives in Bangalore with his wife and three-year-old son. In his spare time, he enjoys playing guitar, coding, reading, and watching soccer.

About the Technical Reviewer

Manoj Patil has worked in the software industry for 19 years. He received an engineering degree from COEP, Pune (India), and has been enjoying his exciting IT journey ever since.

As a principal architect at TatvaSoft, Manoj has taken many initiatives in the organization, ranging from training and mentoring teams, leading data science and ML practice, to successfully designing client solutions from different functional domains.

He began his career as a Java programmer but is fortunate to have worked on multiple frameworks with multiple languages and can claim to be a full stack developer. In the last five years, Manoj has worked extensively in the field of BI, big data, and machine learning, using such technologies as Hitachi Vantara (Pentaho), the Hadoop ecosystem, TensorFlow, Python-based libraries, and more.

He is passionate about learning new technologies, trends, and reviewing books. When he's not working, he's either exercising or reading/listening to infinitheism literature.

Acknowledgments

This is my second book on Spark, and along the way, I have come to realize my love for handling big data and performing machine learning as well. Going forward, I intend to write many more books, but first, let me thank a few people who have helped me along this journey. First, I must thank the most important person in my life, my beloved wife, Neha, who selflessly supported me throughout and sacrificed so much to ensure that I completed this book.

I must thank Celestin Suresh John, who believed in me and extended the opportunity to write this book. Aditee Mirashi is one of the best editors in India. This is my second book with her, and it was even more exciting to work with her this time. As usual, she was extremely supportive and always there to accommodate my requests. I especially would like to thank Jim Markham, who dedicated his time to reading every single chapter and offered so many useful suggestions. Thanks, Jim, I really appreciate your input. I also want to thank Manoj Patil, who had the patience to review every line of code and check the appropriateness of each example. Thank you for your feedback and encouragement. It really made a difference to me and the book.

I also want to thank the many mentors who have constantly forced me to pursue my dreams. Thank you Sebastian Keupers, Dr. Vijay Agneeswaran, Sreenivas Venkatraman, Shoaib Ahmed, and Abhishek Kumar, for your time. Finally, I am infinitely grateful to my son, Ziaan, and my parents, for their endless love and support, irrespective of circumstances. You all make my world beautiful.

Introduction

The idea of writing this book had already been seeded while I was working on my first book, and there was a strong reason for that. The earlier book was more focused on machine learning using big data and essentially did not deep-dive sufficiently into supporting aspects, but this book goes a little deeper into the internals of Spark's machine learning library, as well as analyzing of streaming data. It is a good reference point for someone who wants to learn more about how to automate different workflows and build pipelines to handle real-time data.

This book is divided into three main sections. The first provides an introduction to Spark and data analysis on big data; the middle section discusses using Airflow for executing different jobs, in addition to data analysis on streaming data, using the structured streaming features of Spark. The final section covers translation of a business problem into machine learning and solving it, using Spark's machine learning library, with a deep dive into deep learning as well.

This book might also be useful to data analysts and data engineers, as it covers the steps of big data processing using PySpark. Readers who want to make a transition to the data science and machine learning fields will also find this book a good starting point and can gradually tackle more complicated areas later. The case studies and examples given in the book make it really easy to follow and understand the related fundamental concepts. Moreover, there are very few books available on PySpark, and this book certainly adds value to readers' knowledge. The strength of this book lies in its simplicity and on its application of machine learning to meaningful datasets.

INTRODUCTION

I have tried my best to put all my experience and knowledge into this book, and I feel it is particularly relevant to what businesses are seeking in order to solve real challenges. I hope that it will provide you with some useful takeaways.

CHAPTER 1

Introduction to Spark

As this book is about Spark, it makes perfect sense to start the first chapter by looking into some of Spark's history and its different components. This introductory chapter is divided into three sections. In the first, I go over the evolution of data and how it got as far as it has, in terms of size. I'll touch on three key aspects of data. In the second section, I delve into the internals of Spark and go over the details of its different components, including its architecture and modus operandi. The third and final section of this chapter focuses on how to use Spark in a cloud environment.

History

The birth of the Spark project occurred at the Algorithms, Machine, and People (AMP) Lab at the University of California, Berkeley. The project was initiated to address the potential issues in the Hadoop MapReduce framework. Although Hadoop MapReduce was a groundbreaking framework to handle big data processing, in reality, it still had a lot of limitations in terms of speed. Spark was new and capable of doing in-memory computations, which made it almost 100 times faster than any other big data processing framework. Since then, there has been a continuous increase in adoption of Spark across the globe for big data applications. But before jumping into the specifics of Spark, let's consider a few aspects of data itself.

© Pramod Singh 2019
P. Singh, *Learn PySpark*, https://doi.org/10.1007/978-1-4842-4961-1_1

Data can be viewed from three different angles: the way it is collected, stored, and processed, as shown in Figure 1-1.

Figure 1-1. *Three aspects of data*

Data Collection

A huge shift in the manner in which data is collected has occurred over the last few years. From buying an apple at a grocery store to deleting an app on your mobile phone, every data point is now captured in the back end and collected through various built-in applications. Different Internet of things (IoT) devices capture a wide range of visual and sensory signals every millisecond. It has become relatively convenient for businesses to collect that data from various sources and use it later for improved decision making.

Data Storage

In previous years, no one ever imagined that data would reside at some remote location, or that the cost to store data would be as cheap as it is. Businesses have embraced cloud storage and started to see its benefits over on-premise approaches. However, some businesses still opt for on-premise storage, for various reasons. It's known that data storage began by making use of magnetic tapes. Then the breakthrough introduction of floppy discs made it possible to move data from one place to another. However, the size of the data was still a huge limitation. Flash drives and hard discs made it even easier to store and transfer large amounts of data at a reduced cost. (See Figure 1-2.) The latest trend in the advancement of storage devices has resulted in flash drives capable of storing data up to 2TBs, at a throwaway price.

Figure 1-2. *Evolution of data storage*

This trend clearly indicates that the cost to store data has been reduced significantly over the years and continues to decline. As a result, businesses don't shy away from storing huge amounts of data, irrespective of its kind. From logs to financial and operational transactions to simple employee feedback, everything gets stored.

Data Processing

The final aspect of data is using stored data and processing it for some analysis or to run an application. We have witnessed how efficient computers have become in the last 20 years. What used to take five minutes to execute probably takes less than a second using today's machines with advanced processing units. Hence, it goes without saying that machines can process data much faster and easier. Nonetheless, there is still a limit to the amount of data a single machine can process, regardless of its processing power. So, the underlying idea behind Spark is to use a collection (cluster) of machines and a unified processing engine (Spark) to process and handle huge amounts of data, without compromising on speed and security. This was the ultimate goal that resulted in the birth of Spark.

Spark Architecture

There are five core components that make Spark so powerful and easy to use. The core architecture of Spark consists of the following layers, as shown in Figure 1-3:

- Storage

- Resource management

- Engine

- Ecosystem

- APIs

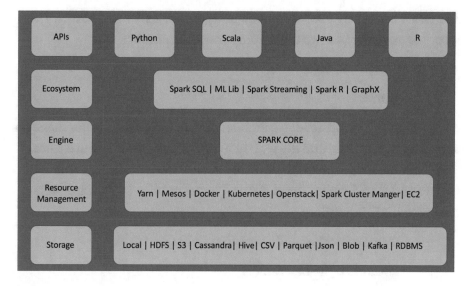

Figure 1-3. *Core components of Spark*

Storage

Before using Spark, data must be made available in order to process it. This data can reside in any kind of database. Spark offers multiple options to use different categories of data sources, to be able to process it on a large scale. Spark allows you to use traditional relational databases as well as NoSQL, such as Cassandra and MongoDB.

Resource Management

The next layer consists of a resource manager. As Spark works on a set of machines (it also can work on a single machine with multiple cores), it is known as a Spark *cluster*. Typically, there is a resource manager in any cluster that efficiently handles the workload between these resources.

The two most widely used resource managers are YARN and Mesos. The resource manager has two main components internally:

1. *Cluster manager*

2. *Worker*

It's kind of like master-slave architecture, in which the cluster manager acts as a master node, and the worker acts as a slave node in the cluster. The cluster manager keeps track of all information pertaining to the worker nodes and their current status. Cluster managers always maintain the following information:

- Status of worker node (busy/available)

- Location of worker node

- Memory of worker node

- Total CPU cores of worker node

The main role of the cluster manager is to manage the worker nodes and assign them tasks, based on the availability and capacity of the worker node. On the other hand, a worker node is only responsible for executing the task it's given by the cluster manager, as shown in Figure 1-4.

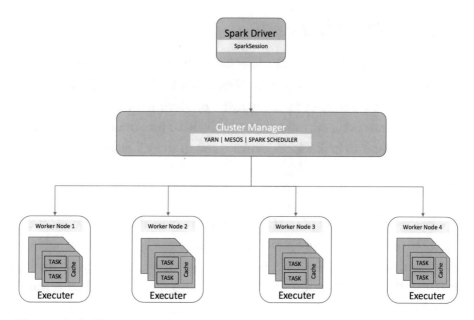

Figure 1-4. *Resource management*

The tasks that are given to the worker nodes are generally the individual pieces of the overall Spark application. The Spark application contains two parts:

1. Task

2. Spark driver

The task is the data processing logic that has been written in either PySpark or Spark R code. It can be as simple as taking a total frequency count of words to a very complex set of instructions on an unstructured dataset. The second component is Spark driver, the main controller of a Spark application, which consistently interacts with a cluster manager to find out which worker nodes can be used to execute the request. The role of the Spark driver is to request the cluster manager to initiate the Spark executor for every worker node.

Engine and Ecosystem

The base of the Spark architecture is its core, which is built on top of RDDs (Resilient Distributed Datasets) and offers multiple APIs for building other libraries and ecosystems by Spark contributors. It contains two parts: the distributed computing infrastructure and the RDD programming abstraction. The default libraries in the Spark toolkit come as four different offerings.

Spark SQL

SQL being used by most of the ETL operators across the globe makes it a logical choice to be part of Spark offerings. It allows Spark users to perform structured data processing by running SQL queries. In actuality, Spark SQL leverages the catalyst optimizer to perform the optimizations during the execution of SQL queries.

Another advantage of using Spark SQL is that it can easily deal with multiple database files and storage systems such as SQL, NoSQL, Parquet, etc.

MLlib

Training machine learning models on big datasets was starting to become a huge challenge, until Spark's MLlib (Machine Learning library) came into existence. MLlib gives you the ability to train machine learning models on huge datasets, using Spark clusters. It allows you to build in supervised, unsupervised, and recommender systems; NLP-based models; and deep learning, as well as within the Spark ML library.

Structured Streaming

The Spark Streaming library provides the functionality to read and process real-time streaming data. The incoming data can be batch data or near real-time data from different sources. Structured Streaming is capable of ingesting real-time data from such sources as Flume, Kafka, Twitter, etc. There is a dedicated chapter on this component later in this book (see Chapter 3).

Graph X

This is a library that sits on top of the Spark core and allows users to process specific types of data (graph dataframes), which consists of nodes and edges. A typical graph is used to model the relationship between the different objects involved. The nodes represent the object, and the edge between the nodes represents the relationship between them. Graph dataframes are mainly used in network analysis, and Graph X makes it possible to have distributed processing of such graph dataframes.

Programming Language APIs

Spark is available in four languages. Because Spark is built using Scala, that becomes the native language. Apart from Scala, we can also use Python, Java, and R, as shown in Figure 1-5.

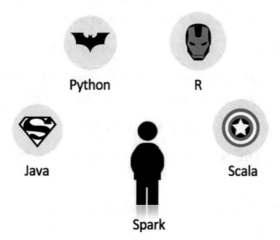

Figure 1-5. *Language APIs*

Setting Up Your Environment

In this final section of this chapter, I will go over how to set up the Spark environment in the cloud. There are multiple ways in which we can use Spark:

- Local setup

- Dockers

- Cloud environment (GCP, AWS, Azure)

- Databricks

Local Setup

It is relatively easy to install and use Spark on a local system, but it fails the core purpose of Spark itself, if it's not used on a cluster. Spark's core offering is distributed data processing, which will always be limited to a local system's capacity, in the case that it's run on a local system, whereas one can benefit more by using Spark on a group of machines instead. However, it is always good practice to have Spark locally, as well as to test code on sample data. So, follow these steps to do so:

1. Ensure that Java is installed; otherwise install Java.

2. Download the latest version of Apache Spark from `https://spark.apache.org/downloads.html`.

3. Extract the files from the zipped folder.

4. Copy all the Spark-related files to their respective directory.

5. Configure the environment variables to be able to run Spark.

6. Verify the installation and run Spark.

Dockers

Another way of using Spark locally is through the containerization technique of dockers. This allows users to wrap all the dependencies and Spark files into a single image, which can be run on any system. We can kill the container after the task is finished and rerun it, if required. To use dockers for running Spark, we must install Docker on the system first and then simply run the following command: [In]: `docker run -it -p 8888:8888 jupyter/pyspark-notebook"`.

Cloud Environments

As discussed earlier in this chapter, for various reasons, local sets are not of much help when it comes to big data, and that's where cloud-based environments make it possible to ingest and process huge datasets in a short period. The real power of Spark can be seen easily while dealing with large datasets (in excess of 100TB). Most of the cloud-based infra-providers allow you to install Spark, which sometimes comes preconfigured as well. One can easily spin up the clusters with required specifications, according to need. One of the cloud-based environments is Databricks.

Databricks

Databricks is a company founded by the creators of Spark, in order to provide the enterprise version of Spark to businesses, in addition to full-fledged support. To increase Spark's adoption among the community and other users, Databricks also provides a free community edition of Spark, with a 6GB cluster (single node). You can increase the size of the cluster by signing up for an enterprise account with Databricks, using the following steps:

1. Search for the Databricks web site and select Databricks Community Edition, as shown in Figure 1-6.

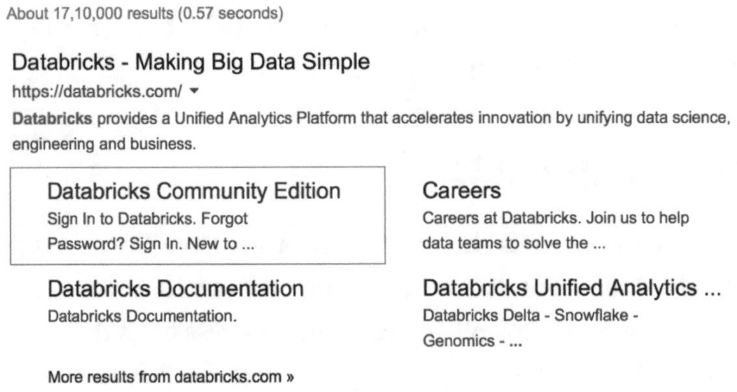

About 17,10,000 results (0.57 seconds)

Databricks - Making Big Data Simple
https://databricks.com/ ▾
Databricks provides a Unified Analytics Platform that accelerates innovation by unifying data science, engineering and business.

Databricks Community Edition	Careers
Sign In to Databricks. Forgot Password? Sign In. New to ...	Careers at Databricks. Join us to help data teams to solve the ...

Databricks Documentation
Databricks Documentation.

Databricks Unified Analytics ...
Databricks Delta - Snowflake - Genomics - ...

More results from databricks.com »

Figure 1-6. *Databricks web page*

2. If you have a user account with Databricks, you can simply log in. If you don't have an account, you must create one, in order to use Databricks, as shown in Figure 1-7.

Figure 1-7. *Databricks login*

3. Once you are on the home page, you can choose to either load a new data source or create a notebook from scratch, as shown in Figure 1-8. In the latter case, you must have the cluster up and running, to be able to use the notebook. Therefore, you must click New Cluster, to spin up the cluster. (Databricks provides a 6GB AWS EMR cluster.)

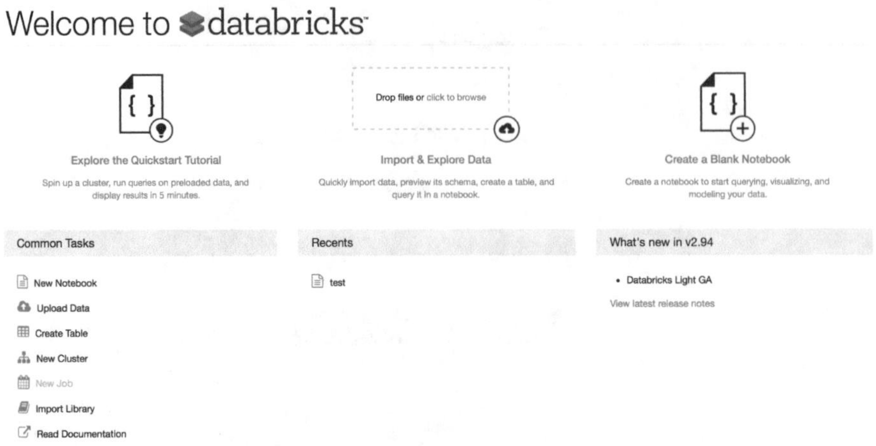

Welcome to ●databricks™

Explore the Quickstart Tutorial

Spin up a cluster, run queries on preloaded data, and display results in 5 minutes.

Drop files or click to browse

Import & Explore Data

Quickly import data, preview its schema, create a table, and query it in a notebook.

Create a Blank Notebook

Create a notebook to start querying, visualizing, and modeling your data.

Common Tasks	Recents	What's new in v2.94
New Notebook	test	• Databricks Light GA
Upload Data		View latest release notes
Create Table		
New Cluster		
New Job		
Import Library		
Read Documentation		

Figure 1-8. *Creating a Databricks notebook*

4. To set up the cluster, you must give a name to the cluster and select the version of Spark that must configure with the Python version, as shown in Figure 1-9. Once all the details are filled in, you must click Create Cluster and wait a couple of minutes, until it spins up.

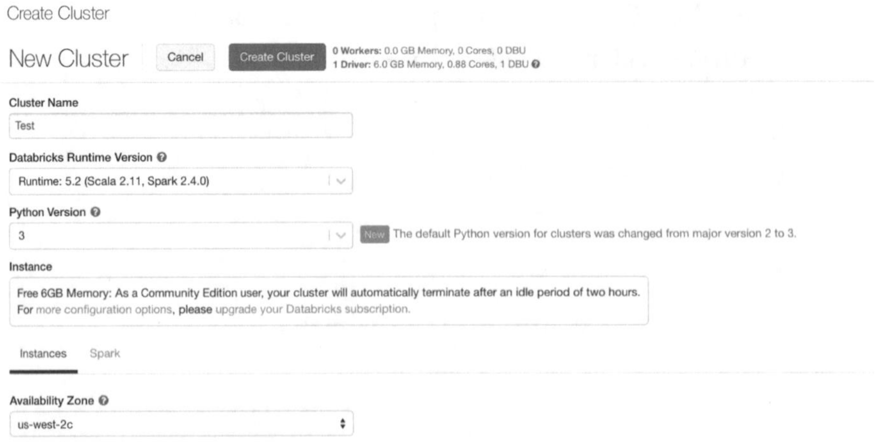

Create Cluster

New Cluster   Cancel   Create Cluster   0 Workers: 0.0 GB Memory, 0 Cores, 0 DBU
1 Driver: 6.0 GB Memory, 0.88 Cores, 1 DBU

Cluster Name

Test

Databricks Runtime Version

Runtime: 5.2 (Scala 2.11, Spark 2.4.0)

Python Version

3   New The default Python version for clusters was changed from major version 2 to 3.

Instance

Free 6GB Memory: As a Community Edition user, your cluster will automatically terminate after an idle period of two hours. For more configuration options, **please** upgrade your Databricks subscription.

Instances   Spark

Availability Zone

us-west-2c

Figure 1-9. *Creating a Databricks cluster*

5. You can also view the status of the cluster by going into the Clusters option on the left side widget, as shown in Figure 1-10. It gives all the information associated with the particular cluster and its current status.

Figure 1-10. *Databricks cluster list*

6. The final step is to open a notebook and attach it to the cluster you just created (Figure 1-11). Once attached, you can start the PySpark code.

Figure 1-11. *Databricks notebook*

Overall, since 2010, when Spark became an open source platform, its users have risen in number consistently, and the community continues to grow every day. It's no surprise that the number of contributors to Spark has outpaced that of Hadoop. Some of the reasons for Spark's popularity were noted in a survey, the results of which are shown in Figure 1-12.

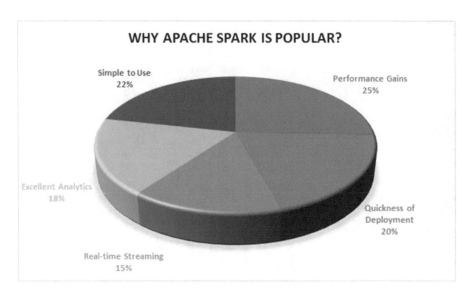

Figure 1-12. *Results of Spark adoption survey*

Conclusion

This chapter provided a brief history of Spark, its core components, and the process of accessing it in a cloud environment. In upcoming chapters, I will delve deeper into the various aspects of Spark and how to build different applications with it.

CHAPTER 2

Data Processing

This chapter covers different steps to preprocess and handle data in PySpark. Preprocessing techniques can certainly vary from case to case, and many different methods can be used to massage the data into desired form. The idea of this chapter is to expose some of the common techniques for dealing with big data in Spark. In this chapter, we are going to go over different steps involved in preprocessing data, such as handling missing values, merging datasets, applying functions, aggregations, and sorting. One major part of data preprocessing is the transformation of numerical columns into categorical ones and vice versa, which we are going to look at over the next few chapters and are based on machine learning. The dataset that we are going to make use of in this chapter is inspired by a primary research dataset and contains a few attributes from the original dataset, with additional columns containing fabricated data points.

Note All the following steps are written in Jupyter Notebook, running Spark on a Docker image (mentioned in Chapter 1). All the subsequent code can also be run in Databricks.

© Pramod Singh 2019
P. Singh, *Learn PySpark*, https://doi.org/10.1007/978-1-4842-4961-1_2

Creating a SparkSession Object

The first step is to create a SparkSession object, in order to use Spark. We also import all the required functions and datatypes from spark.sql:

```
[In]: from pyspark.sql import SparkSession
[In]: spark=SparkSession.builder.appName('data_processing').
getOrCreate()
[In]: import pyspark.sql.functions as F
[In]: from pyspark.sql.types import *
```

Now, instead of directly reading a file to create a dataframe, we go over the process of creating a dataframe, by passing key values. The way we create a dataframe in Spark is by declaring its schema and pass the columns values.

Creating Dataframes

In the following example, we are creating a new dataframe with five columns of certain datatypes (string and integer). As you can see, when we call show on the new dataframe, it is created with three rows and five columns containing the values passed by us.

```
[In]:schema=StructType().add("user_id","string").
add("country","string").add("browser", "string").
add("OS",'string').add("age", "integer")
```

```
[In]: df=spark.createDataFrame([("A203",'India',"Chrome","WIN",
33),("A201",'China',"Safari","MacOS",35),("A205",'UK','Mozilla',
"Linux",25)],schema=schema)
```

```
[In]: df.printSchema()
[Out]:
```

```
root
 |-- user_id: string (nullable = true)
 |-- country: string (nullable = true)
 |-- browser: string (nullable = true)
 |-- OS: string (nullable = true)
 |-- age: integer (nullable = true)
```

```
[In]: df.show()
[Out]:
```

```
+--------+--------+--------+-----+---+
|user_id|country|browser|   OS|age|
+--------+--------+--------+-----+---+
|    A203|   India|  Chrome|  WIN| 33|
|    A201|   China|  Safari|MacOS| 35|
|    A205|      UK|Mozilla|Linux| 25|
+--------+--------+--------+-----+---+
```

Null Values

It is very common to have null values as part of the overall data. Therefore, it becomes critical to add a step to the data processing pipeline, to handle the null values. In Spark, we can deal with null values by either replacing them with some specific value or dropping the rows/columns containing null values.

First, we create a new dataframe (df_na) that contains null values in two of its columns (the schema is the same as in the earlier dataframe). By the first approach to deal with null values, we fill all null values in the present dataframe with a value of 0, which offers a quick fix. We use the fillna function to replace all the null values in the dataframe with 0. By the second approach, we replace the null values in specific columns (country, browser) with 'USA' and 'Safari', respectively.

```
[In]: df_na=spark.createDataFrame([("A203",None,"Chrome","WIN",
33),("A201",'China',None,"MacOS",35),("A205",'UK',"Mozilla",
"Linux",25)],schema=schema)
```

```
[In]: df_na.show()
[Out]:
```

```
+--------+--------+--------+-----+---+
|user_id|country|browser|    OS|age|
+--------+--------+--------+-----+---+
|   A203|   null| Chrome|  WIN| 33|
|   A201|  China|   null|MacOS| 35|
|   A205|     UK|Mozilla|Linux| 25|
+--------+--------+--------+-----+---+
```

```
[In]: df_na.fillna('0').show()
[Out]:
```

```
+--------+--------+--------+-----+---+
|user_id|country|browser|    OS|age|
+--------+--------+--------+-----+---+
|   A203|      0| Chrome|  WIN| 33|
|   A201|  China|      0|MacOS| 35|
|   A205|     UK|Mozilla|Linux| 25|
+--------+--------+--------+-----+---+
```

```
[In]: df_na.fillna({'country':'USA','browser':'Safari'}).show()
[Out]:
```

```
+--------+--------+--------+-----+---+
|user_id|country|browser|    OS|age|
+--------+--------+--------+-----+---+
|   A203|    USA| Chrome|  WIN| 33|
|   A201|  China| Safari|MacOS| 35|
|   A205|     UK|Mozilla|Linux| 25|
+--------+--------+--------+-----+---+
```

In order to drop the rows with any null values, we can simply use the na.drop functionality in PySpark. Whereas if this needs to be done for specific columns, we can pass the set of column names as well, as shown in the following example:

```
[In]: df_na.na.drop().show()
[Out]:
```

```
+-------+-------+-------+-----+---+
|user_id|country|browser|   OS|age|
+-------+-------+-------+-----+---+
|   A205|     UK|Mozilla|Linux| 25|
+-------+-------+-------+-----+---+
```

```
[In]: df_na.na.drop(subset='country').show()
[Out]:
```

```
+-------+-------+-------+-----+---+
|user_id|country|browser|   OS|age|
+-------+-------+-------+-----+---+
|   A201|  China|   null|MacOS| 35|
|   A205|     UK|Mozilla|Linux| 25|
+-------+-------+-------+-----+---+
```

Another very common step in data processing is to replace some data points with particular values. We can use the replace function for this, as shown in the following example. To drop the column of a dataframe, we can use the drop functionality of PySpark.

```
[In]: df_na.replace("Chrome","Google Chrome").show()
[Out]:
```

```
+-------+-------+--------------+-----+---+
|user_id|country|       browser|   OS|age|
+-------+-------+--------------+-----+---+
|   A203|   null|Google Chrome|  WIN| 33|
|   A201|  China|          null|MacOS| 35|
|   A205|     UK|       Mozilla|Linux| 25|
+-------+-------+--------------+-----+---+
```

```
[In]: df_na.drop('user_id').show()
[Out]:
```

```
+-------+-------+-----+---+
|country|browser|   OS|age|
+-------+-------+-----+---+
|  India| Chrome|  WIN| 33|
|  China| Safari|MacOS| 35|
|     UK|Mozilla|Linux| 25|
+-------+-------+-----+---+
```

Now that we have seen how to create a dataframe by passing a value and how to treat missing values, we can create a Spark dataframe, by reading a file (.csv, parquet, etc.). The dataset contains a total of seven columns and 2,000 rows. The summary function allows us to see the statistical measures of the dataset, such as the min, max, and mean of the numerical data present in the dataframe.

```
[In]: df=spark.read.csv("customer_data.csv",header=True,
inferSchema=True)
[In]: df.count()
[Out]: 2000

[In]: len(df.columns)
[Out]: 7

[In]: df.printSchema()
[Out]:

        root
         |-- Customer_subtype: string (nullable = true)
         |-- Number_of_houses: integer (nullable = true)
         |-- Avg_size_household: integer (nullable = true)
         |-- Avg_age: string (nullable = true)
         |-- Customer_main_type: string (nullable = true)
         |-- Avg_Salary: integer (nullable = true)
         |-- label: integer (nullable = true)


[In]: df.show(3)
[Out]:
```

```
+--------------------+----------------+------------------+-----------+--------------------+----------+-----+
|    Customer_subtype|Number_of_houses|Avg_size_household|    Avg_age|  Customer_main_type|Avg_Salary|label|
+--------------------+----------------+------------------+-----------+--------------------+----------+-----+
|Lower class large...|               1|                 3|30-40 years|Family with grown...|     44905|    0|
|Mixed small town ...|               1|                 2|30-40 years|Family with grown...|     37575|    0|
|Mixed small town ...|               1|                 2|30-40 years|Family with grown...|     27915|    0|
+--------------------+----------------+------------------+-----------+--------------------+----------+-----+
only showing top 3 rows
```

```
[In]: df.summary().show()
[Out]:
```

```
+-------+--------------------+----------------+------------------+----------+--------------------+----------------
-+------------------+
|summary|   Customer_subtype| Number_of_houses|Avg_size_household|   Avg_age|  Customer_main_type|        Avg_Salar
y|              label|
+-------+--------------------+----------------+------------------+----------+--------------------+----------------
-+------------------+
|  count|                2000|            2000|              2000|      2000|                2000|             200
0|               2000|
|   mean|                null|          1.1075|            2.6895|      null|                null|      1616908.083
5|             0.0605|
| stddev|                    |            null|0.3873225521186316|0.7914562220841646|    null|            null|6822647.75731214
6|0.2384705099001677|
|    min|Affluent senior a...|               1|                 1|20-30 years|      Average Family|             136
1|                 0|
|    25%|                null|               1|                 2|      null|                null|            2031
5|                 0|
|    50%|                null|               1|                 3|      null|                null|            3142
1|                 0|
|    75%|                null|               1|                 3|      null|                null|            4294
9|                 0|
|    max|  Young, low educated|             10|                 5|70-80 years|Successful hedonists|         4891989
6|                 1|
+-------+--------------------+----------------+------------------+----------+--------------------+----------------
-+------------------+
```

Most of the time, we won't use all the columns present in the dataframe, as some might be redundant and carry very little value in terms of providing useful information. Therefore, subsetting the dataframe becomes critical for having proper data in place for analysis. I'll cover this in the next section.

Subset of a Dataframe

A subset of a dataframe can be created, based on multiple conditions in which we either select a few rows, columns, or data with certain filters in place. In the following examples, you will see how we can create a subset of the original dataframe, based on certain conditions, to demonstrate the process of filtering records.

- Select

- Filter

- Where

Select

In this example, we take one of the dataframe columns, 'Avg_Salary', and create a subset of the original dataframe, using select. We can pass any number of columns that must be present in the subset. We then apply a filter on the dataframe, to extract the records, based on a certain threshold (Avg_Salary > 1000000). Once filtered, we can either take the total count of records present in the subset or take it for further processing.

```
[In]: df.select(['Customer_subtype','Avg_Salary']).show()
[Out]:
```

```
+--------------------+----------+
|    Customer_subtype|Avg_Salary|
+--------------------+----------+
|Lower class large...|     44905|
|Mixed small town ...|     37575|
|Mixed small town ...|     27915|
|Modern, complete ...|     19504|
|   Large family farms|     34943|
|     Young and rising|     13064|
|Large religious f...|     29090|
|Lower class large...|      6895|
|Lower class large...|     35497|
|     Family starters|     30800|
|       Stable family|     39157|
|Modern, complete ...|     40839|
|Lower class large...|     30008|
|        Mixed rurals|     37209|
|     Young and rising|     45361|
|Lower class large...|     45650|
|Traditional families|     18982|
|Mixed apartment d...|     30093|
|Young all america...|     27097|
|Low income catholics|     23511|
+--------------------+----------+
only showing top 20 rows
```

```
[In]: df.filter(df['Avg_Salary'] > 1000000).count()
[In]: 128
[In]: df.filter(df['Avg_Salary'] > 1000000).show()
```

```
+--------------------+----------------+----------------+-----------+--------------------+----------+-----+
|    Customer_subtype|Number_of_houses|Avg_size_household|   Avg_age|   Customer_main_type|Avg_Salary|label|
+--------------------+----------------+----------------+-----------+--------------------+----------+-----+
|  High status seniors|               1|               3|40-50 years|Successful hedonists|   4670288|    0|
|  High status seniors|               1|               3|50-60 years|Successful hedonists|   9561873|    0|
|  High status seniors|               1|               2|40-50 years|Successful hedonists|  18687005|    0|
|  High status seniors|               1|               2|40-50 years|Successful hedonists|  24139960|    0|
|  High status seniors|               1|               2|50-60 years|Successful hedonists|   6718606|    0|
|High Income, expe...|               1|               3|40-50 years|Successful hedonists|  19347139|    0|
|High Income, expe...|               1|               3|40-50 years|Successful hedonists|   5243902|    0|
|  High status seniors|               2|               3|40-50 years|Successful hedonists|   6138618|    0|
|High Income, expe...|               1|               3|50-60 years|Successful hedonists|  24930053|    0|
|  High status seniors|               1|               2|50-60 years|Successful hedonists|  12545905|    1|
|High Income, expe...|               1|               3|40-50 years|Successful hedonists|  29976435|    0|
|  High status seniors|               1|               2|50-60 years|Successful hedonists|  24639614|    0|
|  High status seniors|               1|               2|40-50 years|Successful hedonists|  16073966|    0|
|High Income, expe...|               1|               4|40-50 years|Successful hedonists|  35032441|    1|
|High Income, expe...|               1|               2|50-60 years|Successful hedonists|   8354410|    0|
|  High status seniors|               1|               1|60-70 years|Successful hedonists|  20241068|    0|
|  High status seniors|               1|               1|50-60 years|Successful hedonists|  45592572|    0|
|  High status seniors|               1|               2|50-60 years|Successful hedonists|  10289449|    0|
|High Income, expe...|               1|               2|50-60 years|Successful hedonists|   5586401|    0|
|  High status seniors|               1|               2|50-60 years|Successful hedonists|  41699271|    0|
+--------------------+----------------+----------------+-----------+--------------------+----------+-----+
only showing top 20 rows
```

Filter

We can also apply more than one filter on the dataframe, by including more conditions, as shown following. This can be done in two ways: first, by applying consecutive filters, then by using (&, or) operands with a where statement.

```
[In]: df.filter(df['Avg_Salary'] > 500000).filter(df['Number_
of_houses'] > 2).show()
[Out]:
```

```
+--------------------+----------------+----------------+-----------+--------------------+----------+-----+
|    Customer_subtype|Number_of_houses|Avg_size_household|   Avg_age|   Customer_main_type|Avg_Salary|label|
+--------------------+----------------+----------------+-----------+--------------------+----------+-----+
|Affluent senior a...|               3|               2|50-60 years|Successful hedonists|    596723|    0|
|Affluent senior a...|               3|               2|50-60 years|Successful hedonists|    944444|    0|
|Affluent senior a...|               3|               2|50-60 years|Successful hedonists|    788477|    0|
|Affluent senior a...|               3|               2|50-60 years|Successful hedonists|    994077|    0|
+--------------------+----------------+----------------+-----------+--------------------+----------+-----+
```

Where

```
[In]: df.where((df['Avg_Salary'] > 500000) & (df['Number_of_
houses'] > 2)).show()
```

```
[Out]:
```

```
+--------------------+----------------+------------------+------------+-------------------+----------+-----+
|     Customer_subtype|Number_of_houses|Avg_size_household|     Avg_age|  Customer_main_type|Avg_Salary|label|
+--------------------+----------------+------------------+------------+-------------------+----------+-----+
|Affluent senior a...|               3|                 2|50-60 years|Successful hedonists|    596723|    0|
|Affluent senior a...|               3|                 2|50-60 years|Successful hedonists|    944444|    0|
|Affluent senior a...|               3|                 2|50-60 years|Successful hedonists|    788477|    0|
|Affluent senior a...|               3|                 2|50-60 years|Successful hedonists|    994077|    0|
+--------------------+----------------+------------------+------------+-------------------+----------+-----+
```

Now that we have seen how to create a subset from a dataframe, we can move on to aggregations in PySpark.

Aggregations

Any kind of aggregation can be broken simply into three stages, in the following order:

- Split

- Apply

- Combine

The first step is to split the data, based on a column or group of columns, followed by performing the operation on those small individual groups (count, max, avg, etc.). Once the results are in for each set of groups, the last step is to combine all these results.

In the following example, we aggregate the data, based on 'Customer subtype', and simply count the number of records in each category. We use the groupBy function in PySpark. The output of this is not in any particular order, as we have not applied any sorting to the results. Therefore, we will also see how we can apply any type of sorting to the final results. Because we have seven columns in the dataframe—all are categorical columns except for one (Avg_Salary), we can iterate over each column and apply aggregation as in the following example:

```
[In]: df.groupBy('Customer_subtype').count().show()
[Out]:
```

```
+--------------------+-----+
|    Customer_subtype|count|
+--------------------+-----+
|Large family, emp...|   56|
|Religious elderly...|   47|
|Large religious f...|  107|
|Modern, complete ...|   93|
|    Village families|   68|
|Young all america...|   62|
|Young urban have-...|    4|
|Young seniors in ...|   22|
|Fresh masters in ...|    2|
|High Income, expe...|   52|
|Lower class large...|  288|
| Residential elderly|    6|
|Senior cosmopolitans|    1|
|        Mixed rurals|   67|
|Career and childcare|   33|
|Low income catholics|   72|
|Mixed apartment d...|   34|
|Seniors in apartm...|   17|
|Middle class fami...|  122|
|Traditional families|  129|
+--------------------+-----+
only showing top 20 rows
```

```
[In]:
for col in df.columns:
    if col !='Avg_Salary':
        print(f" Aggregation for  {col}")
            df.groupBy(col).count().orderBy('count',ascending=
            False).show(truncate=False)
```

[Out]:

```
          *** Aggregation for  Customer_subtype ***
+-------------------------------------------------+-----+
|Customer_subtype                                 |count|
+-------------------------------------------------+-----+
|Lower class large families                       |288  |
|Traditional families                             |129  |
|Middle class families                            |122  |
|Large religious families                         |107  |
|Modern, complete families                        |93   |
|Couples with teens 'Married with children'       |83   |
|Young and rising                                 |78   |
|High status seniors                              |76   |
|Low income catholics                             |72   |
|Mixed seniors                                    |71   |
|Village families                                 |68   |
|Mixed rurals                                     |67   |
|Young all american family                        |62   |
|Stable family                                    |62   |
|Young, low educated                              |56   |
|Large family, employed child                     |56   |
|Family starters                                  |55   |
|High Income, expensive child                     |52   |
|Mixed small town dwellers                        |47   |
|Religious elderly singles                        |47   |
+-------------------------------------------------+-----+
only showing top 20 rows

          *** Aggregation for  Number_of_houses ***
+----------------+-----+
|Number_of_houses|count|
+----------------+-----+
|1               |1808 |
|2               |178  |
|3               |12   |
|5               |1    |
|10              |1    |
+----------------+-----+

          *** Aggregation for  Avg_size_household ***
+------------------+-----+
|Avg_size_household|count|
+------------------+-----+
|3                 |900  |
|2                 |730  |
|4                 |255  |
|1                 |94   |
|5                 |21   |
+------------------+-----+
```

```
*** Aggregation for  Avg_age ***
+-----------+-----+
|Avg_age    |count|
+-----------+-----+
|40-50 years|1028 |
|30-40 years|496  |
|50-60 years|373  |
|60-70 years|64   |
|20-30 years|31   |
|70-80 years|8    |
+-----------+-----+

*** Aggregation for  Customer_main_type ***
+---------------------+-----+
|Customer_main_type   |count|
+---------------------+-----+
|Family with grown ups|542  |
|Average Family       |308  |
|Conservative families|236  |
|Retired and Religious|202  |
|Successful hedonists |194  |
|Living well          |178  |
|Driven Growers       |172  |
|Farmers              |93   |
|Cruising Seniors     |60   |
|Career Loners        |15   |
+---------------------+-----+

*** Aggregation for  label ***
+-----+-----+
|label|count|
+-----+-----+
|0    |1879 |
|1    |121  |
+-----+-----+
```

As mentioned, we can have different kinds of operations on groups of records, such as

- Mean

- Max

- Min

- Sum

The following examples cover some of these, based on different groupings. F refers to the Spark `sql` function here.

```
[In]: df.groupBy('Customer_main_type').agg(F.mean('Avg_
Salary')).show()
[Out]:
```

```
+--------------------+--------------------+
|  Customer_main_type|     avg(Avg_Salary)|
+--------------------+--------------------+
|             Farmers|   30209.333333333332|
|       Career Loners|             32272.6|
|  Retired and Relig...|    27338.80693069307|
|  Successful hedonists|1.6278923510309279E7|
|         Living well|   31194.044943820223|
|      Average Family|   104256.62337662338|
|     Cruising Seniors|   28870.333333333332|
|  Conservative fami...|   29504.419491525423|
|      Driven Growers|    30769.04069767442|
|  Family with grown...|   28114.191881918818|
+--------------------+--------------------+
```

```
[In]: df.groupBy('Customer_main_type').agg(F.max('Avg_
Salary')).show()
[Out]:
```

```
+--------------------+---------------+
|  Customer_main_type|max(Avg_Salary)|
+--------------------+---------------+
|             Farmers|          49965|
|       Career Loners|          49903|
|  Retired and Relig...|          49564|
|  Successful hedonists|       48919896|
|         Living well|          49816|
|      Average Family|         991838|
|     Cruising Seniors|          49526|
|  Conservative fami...|          49965|
|      Driven Growers|          49932|
|  Family with grown...|          49901|
+--------------------+---------------+
```

```
[In]: df.groupBy('Customer_main_type').agg(F.min('Avg_
Salary')).show()
[Out]:
```

```
+--------------------+---------------+
| Customer_main_type|min(Avg_Salary)|
+--------------------+---------------+
|             Farmers|          10469|
|       Career Loners|          13246|
|Retired and Relig...|           1361|
|Successful hedonists|          12705|
|         Living well|          10418|
|      Average Family|          10506|
|     Cruising Seniors|          10100|
|Conservative fami...|          10179|
|      Driven Growers|          10257|
|Family with grown...|           1502|
+--------------------+---------------+
```

```
[In]: df.groupBy('Customer_main_type').agg(F.sum('Avg_
Salary')).show()
[Out]:
```

```
+--------------------+---------------+
| Customer_main_type|sum(Avg_Salary)|
+--------------------+---------------+
|             Farmers|        2809468|
|       Career Loners|         484089|
|Retired and Relig...|        5522439|
|Successful hedonists|     3158111161|
|         Living well|        5552540|
|      Average Family|       32111040|
|     Cruising Seniors|        1732220|
|Conservative fami...|        6963043|
|      Driven Growers|        5292275|
|Family with grown...|       15237892|
+--------------------+---------------+
```

Sometimes, there is simply a need to sort the data with aggregation or without any sort of aggregation. That's where we can make use of the 'sort' and 'orderBy' functionality of PySpark, to rearrange data in a particular order, as shown in the following examples:

```
[In]: df.sort("Avg_Salary", ascending=False).show()
[Out]:
```

Customer_subtype	Number_of_houses	Avg_size_household	Avg_age	Customer_main_type	Avg_Salary	label
High status seniors	1	2	60-70 years	Successful hedonists	48919896	0
High Income, expe...	1	2	50-60 years	Successful hedonists	48177970	0
High Income, expe...	1	2	50-60 years	Successful hedonists	48069548	1
High Income, expe...	1	3	40-50 years	Successful hedonists	46911924	0
High status seniors	1	3	40-50 years	Successful hedonists	46614009	0
High Income, expe...	1	3	30-40 years	Successful hedonists	45952441	0
High Income, expe...	1	3	40-50 years	Successful hedonists	45864609	1
High status seniors	1	1	50-60 years	Successful hedonists	45592572	0
High status seniors	1	2	50-60 years	Successful hedonists	45170899	0
High Income, expe...	1	3	50-60 years	Successful hedonists	44843830	0
High status seniors	1	2	50-60 years	Successful hedonists	43230349	0
High status seniors	1	5	40-50 years	Successful hedonists	43181830	0
High status seniors	1	1	30-40 years	Successful hedonists	42631926	0
High status seniors	1	2	50-60 years	Successful hedonists	41919020	0
High status seniors	1	2	50-60 years	Successful hedonists	41699271	0
High status seniors	1	2	40-50 years	Successful hedonists	41398953	0
High Income, expe...	1	3	40-50 years	Successful hedonists	41269615	1
High status seniors	1	2	50-60 years	Successful hedonists	41192397	1
High status seniors	1	2	40-50 years	Successful hedonists	40564335	0
High status seniors	1	2	50-60 years	Successful hedonists	40453887	0

only showing top 20 rows

```
[In]: df.groupBy('Customer_subtype').agg(F.avg('Avg_Salary').
alias('mean_salary')).orderBy('mean_salary',ascending=False).
show(50,False)
[Out]:
```

```
+--------------------------------------------+--------------------+
|Customer_subtype                            |mean_salary         |
+--------------------------------------------+--------------------+
|High status seniors                         |2.507677857894737E7 |
|High Income, expensive child                |2.3839817807692308E7|
|Affluent young families                     |662068.7777777778   |
|Affluent senior apartments                  |653638.8235294118   |
|Senior cosmopolitans                        |49903.0             |
|Students in apartments                      |35532.142857142855  |
|Large family farms                          |33135.61538461538   |
|Young, low educated                         |33072.21428571428   |
|Large family, employed child                |32867.857142857145  |
|Suburban youth                              |32558.0             |
|Village families                            |32449.470588235294  |
|Middle class families                       |31579.385245901638  |
|Modern, complete families                   |31576.0             |
|Etnically diverse                           |31572.0             |
|Young and rising                            |30795.897435897437  |
|Mixed seniors                               |30759.267605633802  |
|Very Important Provincials                  |30548.40625         |
|Religious elderly singles                   |30540.59574468085   |
|Family starters                             |30376.2             |
|Career and childcare                        |30110.939393939392  |
|Young seniors in the city                   |30105.136363636364  |
|Seniors in apartments                       |30090.882352941175  |
|Large religious families                    |29652.196261682242  |
|Stable family                               |29619.032258064515  |
|Mixed apartment dwellers                    |29587.647058823528  |
|Young all american family                   |29563.3064516129    |
|Porchless seniors: no front yard            |29509.827586206895  |
|Traditional families                        |29381.84496124031   |
|Mixed rurals                                |29073.761194029852  |
|Mixed small town dwellers                   |28982.106382978724  |
|Residential elderly                         |28866.166666666668  |
|Couples with teens 'Married with children'  |28155.807228915663  |
|Fresh masters in the city                   |27645.0             |
|Dinki's (double income no kids)             |26231.117647058825  |
|Lower class large families                  |26012.628472222223  |
|Young urban have-nots                       |25751.0             |
|Own home elderly                            |25677.666666666668  |
|Single youth                                |24403.25            |
|Low income catholics                        |21713.777777777777  |
+--------------------------------------------+--------------------+
```

```
[In]: df.groupBy('Customer_subtype').agg(F.max('Avg_Salary').
alias('max_salary')).orderBy('max_salary',ascending=False).
show()
[Out]:
```

```
+--------------------+----------+
|    Customer_subtype|max_salary|
+--------------------+----------+
| High status seniors|  48919896|
|High Income, expe...|  48177970|
|Affluent senior a...|    994077|
|Affluent young fa...|    991838|
|   Large family farms|     49965|
|Traditional families|     49965|
|Middle class fami...|     49932|
|Senior cosmopolitans|     49903|
|Mixed small town ...|     49901|
|Lower class large...|     49899|
|        Mixed seniors|     49876|
|     Young and rising|     49816|
|        Mixed rurals|     49785|
|Modern, complete ...|     49729|
| Young, low educated|     49626|
|Mixed apartment d...|     49621|
|      Family starters|     49602|
|     Village families|     49575|
|Religious elderly...|     49564|
|        Stable family|     49548|
+--------------------+----------+
only showing top 20 rows
```

In some cases, we must also collect the list of values for particular groups or for individual categories. For example, let's say a customer goes to an online store and accesses different pages on the store's web site. If we have to collect all the customer's activities in a list, we can use the collect functionality in PySpark. We can collect values in two different ways:

- Collect List

- Collect Set

Collect

Collect list provides all the values in the original order of occurrence (they can be reversed as well), and collect set provides only the unique values, as shown in the following example. We consider grouping on Customer subtype and collecting the Numberof houses values in a new column, using list and set separately.

```
[In]: df.groupby("Customer_subtype").agg(F.collect_set("Number_
of_houses")).show()
[Out]:
```

```
+--------------------+----------------------------+
|    Customer_subtype|collect_set(Number_of_houses)|
+--------------------+----------------------------+
|Large family, emp...|                      [1, 2]|
|Religious elderly...|                      [1, 2]|
|Large religious f...|                      [1, 2]|
|Modern, complete ...|                      [1, 2]|
|    Village families|                      [1, 2]|
|Young all america...|                      [1, 2]|
|Young urban have-...|                      [1, 2]|
|Young seniors in ...|                   [1, 2, 3]|
|Fresh masters in ...|                         [1]|
|High Income, expe...|                         [1]|
|Lower class large...|                      [1, 2]|
|  Residential elderly|                   [1, 2, 3]|
|Senior cosmopolitans|                         [3]|
|        Mixed rurals|                         [1]|
|Career and childcare|                      [1, 2]|
|Low income catholics|                         [1]|
|Mixed apartment d...|                   [1, 2, 3]|
|Seniors in apartm...|                      [1, 2]|
|Middle class fami...|                      [1, 2]|
|Traditional families|                      [1, 2]|
+--------------------+----------------------------+
only showing top 20 rows
```

```
[In]:
df.groupby("Customer_subtype").agg(F.collect_list("Number_of_
houses")).show()
[Out]:
```

```
+--------------------+-------------------------------+
|     Customer_subtype|collect_list(Number_of_houses)|
+--------------------+-------------------------------+
|Large family, emp...|        [2, 1, 2, 1, 2, 1...|
|Religious elderly...|        [1, 1, 1, 1, 1, 1...|
|Large religious f...|        [2, 1, 1, 2, 1, 1...|
|Modern, complete ...|        [1, 1, 2, 1, 1, 1...|
|    Village families|        [1, 1, 1, 1, 1, 1...|
|Young all america...|        [1, 1, 2, 2, 1, 1...|
|Young urban have-...|               [1, 2, 1, 1]|
|Young seniors in ...|        [1, 1, 1, 1, 1, 2...|
|Fresh masters in ...|                   [1, 1]|
|High Income, expe...|        [1, 1, 1, 1, 1, 1...|
|Lower class large...|        [1, 1, 1, 1, 1, 1...|
| Residential elderly|        [3, 1, 1, 3, 2, 1]|
| Senior cosmopolitans|                     [3]|
|        Mixed rurals|        [1, 1, 1, 1, 1, 1...|
|Career and childcare|        [2, 1, 1, 1, 1, 1...|
|Low income catholics|        [1, 1, 1, 1, 1, 1...|
|Mixed apartment d...|        [2, 3, 1, 1, 1, 1...|
|Seniors in apartm...|        [2, 2, 2, 2, 1, 2...|
|Middle class fami...|        [1, 1, 2, 1, 1, 1...|
|Traditional families|        [1, 1, 1, 1, 1, 1...|
+--------------------+-------------------------------+
only showing top 20 rows
```

The need to create a new column with a constant value can be very common. Therefore, we can do that in PySpark, using the 'lit' function. In the following example, we create a new column with a constant value:

```
[In]: df=df.withColumn('constant',F.lit('finance'))
[In]: df.select('Customer_subtype','constant').show()
```

`[Out]:`

```
+--------------------+--------+
|    Customer_subtype|constant|
+--------------------+--------+
|Lower class large...| finance|
|Mixed small town ...| finance|
|Mixed small town ...| finance|
|Modern, complete ...| finance|
|   Large family farms| finance|
|      Young and rising| finance|
|Large religious f...| finance|
|Lower class large...| finance|
|Lower class large...| finance|
|      Family starters| finance|
|         Stable family| finance|
|Modern, complete ...| finance|
|Lower class large...| finance|
|         Mixed rurals| finance|
|      Young and rising| finance|
|Lower class large...| finance|
|Traditional families| finance|
|Mixed apartment d...| finance|
|Young all america...| finance|
|Low income catholics| finance|
+--------------------+--------+
only showing top 20 rows
```

Because we are dealing with dataframes, it is a common requirement to apply certain custom functions on specific columns and get the output. Hence, we make use of UDFs, in order to apply Python functions on one or more columns.

User-Defined Functions (UDFs)

In this example, we are trying to name the age categories and create a standard Python function (age_category) for the same. In order to apply this on the Spark dataframe, we create a UDF object, using this Python function. The only requirement is to mention the return type of the function. In this case, it is simply a string value.

```
[In]: from pyspark.sql.functions import udf
[In]: df.groupby("Avg_age").count().show()
[Out]:
```

```
+-----------+-----+
|    Avg_age|count|
+-----------+-----+
|70-80 years|    8|
|50-60 years|  373|
|30-40 years|  496|
|20-30 years|   31|
|60-70 years|   64|
|40-50 years| 1028|
+-----------+-----+
```

```
[In]: def age_category(age):
    if age  == "20-30 years":
        return "Young"
    elif age== "30-40 years":
        return "Mid Aged"
    elif ((age== "40-50 years") or (age== "50-60 years")) :
        return "Old"
    else:
        return "Very Old"
```

```
[In]: age_udf=udf(age_category,StringType())
[In]: df=df.withColumn('age_category',age_udf(df['Avg_age']))
[In]: df.select('Avg_age','age_category').show()
[Out]:
```

```
+-----------+------------+
|   Avg_age|age_category|
+-----------+------------+
|30-40 years|    Mid Aged|
|30-40 years|    Mid Aged|
|30-40 years|    Mid Aged|
|40-50 years|         Old|
|30-40 years|    Mid Aged|
|20-30 years|       Young|
|30-40 years|    Mid Aged|
|40-50 years|         Old|
|50-60 years|         Old|
|40-50 years|         Old|
|40-50 years|         Old|
|40-50 years|         Old|
|40-50 years|         Old|
|40-50 years|         Old|
|30-40 years|    Mid Aged|
|40-50 years|         Old|
|40-50 years|         Old|
|40-50 years|         Old|
|30-40 years|    Mid Aged|
|50-60 years|         Old|
+-----------+------------+
only showing top 20 rows
```

```
[In]: df.groupby("age_category").count().show()
[Out]:
```

```
+------------+-----+
|age_category|count|
+------------+-----+
|    Mid Aged|  496|
|    Very Old|   72|
|         Old| 1401|
|       Young|   31|
+------------+-----+
```

Pandas UDFs are another recent advancement, so let's review them now.

Pandas UDF

Pandas UDFs are much faster and efficient, in terms of processing and execution time, compared to standard Python UDFs. The main difference between a normal Python UDF and a Pandas UDF is that a Python UDF is executed row by row and, therefore, really doesn't offer the advantage of a distributed framework. It can take longer, compared to a Pandas UDF, which executes block by block and gives faster results. There are three different types of Pandas UDFs: scalar, grouped map, and grouped agg. The only difference in using a Pandas UDF compared to a traditional UDF lies in the declaration. In the following example, we try to scale the Avg_Salary values by applying scaling. We first take the min and max values of Avg_Salary, subtract from each value the minimum salary from each value, and then divide by the difference between max and min.

$$\frac{X - X_{min}}{X_{max} - X_{min}}$$

```
[In]: df.select('Avg_Salary').summary().show()
[Out]:
```

summary	Avg_Salary
count	2000
mean	1616908.0835
stddev	6822647.757312146
min	1361
25%	20315
50%	31421
75%	42949
max	48919896

```
[In]: min_sal=1361
[In]: max_sal=48919896
```

```
[In]: from pyspark.sql.functions import pandas_udf,
PandasUDFType

[In]: def scaled_salary(salary):
        scaled_sal=(salary-min_sal)/(max_sal-min_sal)
        return scaled_sal

[In]: scaling_udf = pandas_udf(scaled_salary, DoubleType())
[In]:df.withColumn("scaled_salary",scaling_udf(df['Avg_
Salary'])).show(10,False)
[Out]:
```

```
+---------------------------+-----------------+------------------+----------+---------------------+-----------+-----+--
------+--------------------+
|Customer_subtype           |Number_of_houses|Avg_size_household|Avg_age   |Customer_main_type   |Avg_Salary|label|co
nstant|scaled_salary       |
+---------------------------+-----------------+------------------+----------+---------------------+-----------+-----+--
------+--------------------+
|Lower class large families|1               |3                 |30-40 years|Family with grown ups|44905     |0    |fi
nance |8.901329526732557E-4 |
|Mixed small town dwellers |1               |2                 |30-40 years|Family with grown ups|37575     |0    |fi
nance |7.40291997705982E-4  |
|Mixed small town dwellers |1               |2                 |30-40 years|Family with grown ups|27915     |0    |fi
nance |5.42820834679534E-4  |
|Modern, complete families |1               |3                 |40-50 years|Average Family       |19504     |0    |fi
nance |3.708819162307293E-4 |
|Large family farms         |1               |4                 |30-40 years|Farmers              |34943     |0    |fi
nance |6.864882605335584E-4 |
|Young and rising           |1               |2                 |20-30 years|Living well          |13064     |0    |fi
nance |2.3923447421309735E-4|
|Large religious families  |2               |3                 |30-40 years|Conservative families|29090     |0    |fi
nance |5.668403602029373E-4 |
|Lower class large families|1               |2                 |40-50 years|Family with grown ups|6895      |0    |fi
nance |1.1312685467788436E-4|
|Lower class large families|1               |2                 |50-60 years|Family with grown ups|35497     |0    |fi
nance |6.978132112909759E-4 |
|Family starters            |2               |3                 |40-50 years|Average Family       |30800     |0    |fi
nance |6.017964356455073E-4 |
+---------------------------+-----------------+------------------+----------+---------------------+-----------+-----+--
------+--------------------+
only showing top 10 rows
```

This is how we can use both conventional and Pandas UDFs to apply different conditions on the dataframe, as required.

Joins

Merging different datasets is a very generic requirement present in most of data-processing pipelines in the big data world. PySpark offers a very convenient way to merge and pivot your dataframe values, as required. In the following example, we create a fabricated dataframe with some dummy Region Code values for all Customer types. The idea is to combine this dataframe with the original dataframe, so as to have these region codes as part of the original dataframe, as a column.

```
[In]: region_data = spark.createDataFrame([('Family with grown
ups','PN'),
                    ('Driven Growers','GJ'),
                    ('Conservative families','DD'),
                    ('Cruising Seniors','DL'),
                    ('Average Family ','MN'),
                    ('Living well','KA'),
                    ('Successful hedonists','JH'),
                    ('Retired and Religious','AX'),
                    ('Career Loners','HY'),('Farmers','JH')],
                    schema=StructType().add("Customer_main_
                    type","string").add("Region Code","string"))
[In]: region_data.show()
[Out]:
```

```
+--------------------+-----------+
| Customer_main_type|Region Code|
+--------------------+-----------+
|Family with grown...|         PN|
|      Driven Growers|         GJ|
|Conservative fami...|         DD|
|    Cruising Seniors|         DL|
|     Average Family |         MN|
|         Living well|         KA|
|Successful hedonists|         JH|
|Retired and Relig...|         AX|
|       Career Loners|         HY|
|             Farmers|         JH|
+--------------------+-----------+
```

```
[In]: new_df=df.join(region_data,on='Customer_main_type')
```

```
[In]: new_df.groupby("Region Code").count().show()
```

[Out]:

```
+------------+-----+
|Region Code|count|
+------------+-----+
|JH          |287  |
|HY          |15   |
|DD          |236  |
|DL          |60   |
|GJ          |172  |
|PN          |542  |
|KA          |178  |
|AX          |202  |
+------------+-----+
```

We took the regional count after joining the original dataframe (df) with the newly created region_data dataframe on the Customer_main_type column.

Pivoting

We can use the pivot function in PySpark to simply create a pivot view of the dataframe for specific columns, as shown in the following example. Here, we are grouping data, based on customer type. Columns represent different age groups. The values inside the pivot table are the sum of the Avg Salary of each of these customer type categories for a particular age group. We also ensure that there are no nulls or empty values, by filling all nulls with 0. In the subsequent example, we create one more pivot table, using the label column and take the sum of Avg Salary as the values inside it.

```
[In]:df.groupBy('Customer_main_type').pivot('Avg_age').
sum('Avg_Salary').fillna(0).show()
[Out]:
```

```
+--------------------+-----------+-----------+-----------+-----------+-----------+-----------+
| Customer_main_type|20-30 years|30-40 years|40-50 years|50-60 years|60-70 years|70-80 years|
+--------------------+-----------+-----------+-----------+-----------+-----------+-----------+
|             Farmers|          0|     462027|    2031235|     316206|          0|          0|
|       Career Loners|     143998|     176639|      25701|     105193|      32558|          0|
|Retired and Relig...|     126350|     336631|    2975266|    1687711|     335357|      61124|
| Successful hedonists|     42261|  171278764| 1223362814| 1563071675|  200340129|      15518|
|         Living well|     460528|    2965303|    1795405|     331304|          0|          0|
|      Average Family|          0|   23682805|    7789464|     412490|     226281|          0|
|     Cruising Seniors|         0|      43302|     303601|     529354|     716425|     139538|
| Conservative fami...|     69390|    2381485|    3449782|     915954|     146432|          0|
|      Driven Growers|          0|    1376260|    3407807|     424272|      83936|          0|
|Family with grown...|     16406|    2620378|    9132414|    3295378|     162820|      10496|
+--------------------+-----------+-----------+-----------+-----------+-----------+-----------+
```

```
[In]:df.groupBy('Customer_main_type').pivot('label').sum('Avg_
Salary').fillna(0).show()
[Out]:
```

```
+--------------------+----------+---------+
|  Customer_main_type|         0|        1|
+--------------------+----------+---------+
|             Farmers|   2734832|    74636|
|       Career Loners|    484089|        0|
|Retired and Relig...|   5328410|   194029|
| Successful hedonists|2720381462|437729699|
|         Living well|   5453384|    99156|
|      Average Family|  26036999|  6074041|
|     Cruising Seniors|   1675841|    56379|
| Conservative fami...|   6595027|   368016|
|      Driven Growers|   4492465|   799810|
|Family with grown...|  14394094|   843798|
+--------------------+----------+---------+
```

We split the data, based on the Customer_main_type column, and took
the cumulative sum of the Avg_Salary of each of the label values (0,1),
using the pivot function.

Window Functions or Windowed Aggregates

This functionality in PySpark allows you to perform certain operations on
groups of records known as "within the window." It calculates the results
for each row within the window. A classic example of using window is the
various aggregations for a user during different sessions. A visitor might

have multiple sessions on a particular web site and, hence `window` can be used to count the total activities of the user during each session. PySpark supports three types of `window` functions:

- Aggregations

- Ranking

- Analytics

In the following example, we import the `window` function, in addition to others, such as `row_number`. The next step is to define the window. Sometimes it can be simply an ordered column, or sometimes it can be based on particular categories within a column. We will see examples of each of these. In the first example, we define the window, which is just based on the sorted `Avg Salary` column, and we rank these salaries. We create a new column `'rank'` and assign ranks to each of the `Avg Salary` values.

```
[In]: from pyspark.sql.window import Window
[In]: from pyspark.sql.functions import col,row_number

[In]: win = Window.orderBy(df['Avg_Salary'].desc())
[In]: df=df.withColumn('rank', row_number().over(win).
alias('rank'))
[In]: df.show()
[Out]:
```

```
+--------------------+----------------+-----------------+-----------+-------------------+----------+-----+----+
|     Customer_subtype|Number_of_houses|Avg_size_household|    Avg_age|  Customer_main_type|Avg_Salary|label|rank|
+--------------------+----------------+-----------------+-----------+-------------------+----------+-----+----+
|  High status seniors|               1|                2|60-70 years|Successful hedonists|  48919896|    0|   1|
|High Income, expe...|               1|                2|50-60 years|Successful hedonists|  48177970|    0|   2|
|High Income, expe...|               1|                2|50-60 years|Successful hedonists|  48069548|    1|   3|
|High Income, expe...|               1|                3|40-50 years|Successful hedonists|  46911924|    0|   4|
|  High status seniors|               1|                3|40-50 years|Successful hedonists|  46614009|    0|   5|
|High Income, expe...|               1|                3|30-40 years|Successful hedonists|  45952441|    0|   6|
|High Income, expe...|               1|                3|40-50 years|Successful hedonists|  45864609|    1|   7|
|  High status seniors|               1|                1|50-60 years|Successful hedonists|  45592572|    0|   8|
|  High status seniors|               1|                2|50-60 years|Successful hedonists|  45170899|    0|   9|
|High Income, expe...|               1|                3|50-60 years|Successful hedonists|  44843830|    0|  10|
|  High status seniors|               1|                2|50-60 years|Successful hedonists|  43230349|    0|  11|
|  High status seniors|               1|                5|40-50 years|Successful hedonists|  43181830|    0|  12|
|  High status seniors|               1|                1|30-40 years|Successful hedonists|  42631926|    0|  13|
|  High status seniors|               1|                2|50-60 years|Successful hedonists|  41919020|    0|  14|
|  High status seniors|               1|                2|50-60 years|Successful hedonists|  41699271|    0|  15|
|  High status seniors|               1|                2|40-50 years|Successful hedonists|  41398953|    0|  16|
|High Income, expe...|               1|                3|40-50 years|Successful hedonists|  41269615|    1|  17|
|  High status seniors|               1|                2|50-60 years|Successful hedonists|  41192397|    1|  18|
|  High status seniors|               1|                2|40-50 years|Successful hedonists|  40564335|    0|  19|
|  High status seniors|               1|                2|50-60 years|Successful hedonists|  40453887|    0|  20|
+--------------------+----------------+-----------------+-----------+-------------------+----------+-----+----+
only showing top 20 rows
```

45

One common requirement is to find the top-three values from a category. In this case, window can be used to get the results. In the following example, we define the window and partition by the Customer subtype column. Basically, what it does is sort the Avg Salary for each of the Customer subtype category, so now we can use filter to fetch the top-three salary values for each group.

```
[In]:win_1=Window.partitionBy("Customer_subtype").
orderBy(df['Avg_Salary'].desc())
[In]: df=df.withColumn('rank', row_number().over(win_1).
alias('rank'))
```

Now that we have a new column rank that consists of the rank or each category of Customer_subtype, we can filter the top-three ranks for each category easily.

```
[In]: df.groupBy('rank').count().orderBy('rank').show()
[Out]:
```

```
+----+-----+
|rank|count|
+----+-----+
|   1|   39|
|   2|   37|
|   3|   36|
|   4|   36|
|   5|   34|
|   6|   34|
|   7|   32|
|   8|   31|
|   9|   31|
|  10|   31|
|  11|   31|
|  12|   31|
|  13|   31|
|  14|   31|
|  15|   31|
|  16|   30|
|  17|   30|
|  18|   27|
|  19|   27|
|  20|   27|
+----+-----+
only showing top 20 rows
```

```
[In]: df.filter(col('rank') < 4).show()
[Out]:
```

```
+--------------------+----------------+------------------+-----------+--------------------+----------+-----+----+
|    Customer_subtype|Number_of_houses|Avg_size_household|    Avg_age|  Customer_main_type|Avg_Salary|label|rank|
+--------------------+----------------+------------------+-----------+--------------------+----------+-----+----+
|Large family, emp...|               2|                 3|30-40 years|Family with grown...|     49418|    0|   1|
|Large family, emp...|               1|                 4|40-50 years|Family with grown...|     48390|    0|   2|
|Large family, emp...|               1|                 3|40-50 years|Family with grown...|     48272|    0|   3|
|Religious elderly...|               1|                 2|50-60 years|Retired and Relig...|     49564|    0|   1|
|Religious elderly...|               1|                 2|60-70 years|Retired and Relig...|     49373|    0|   2|
|Religious elderly...|               1|                 1|70-80 years|Retired and Relig...|     49059|    0|   3|
|Large religious f...|               1|                 3|40-50 years|Conservative fami...|     49547|    0|   1|
|Large religious f...|               1|                 2|60-70 years|Conservative fami...|     49268|    0|   2|
|Large religious f...|               1|                 3|40-50 years|Conservative fami...|     49040|    0|   3|
|Modern, complete ...|               1|                 3|30-40 years|      Average Family|     49729|    0|   1|
|Modern, complete ...|               1|                 4|30-40 years|      Average Family|     48801|    0|   2|
|Modern, complete ...|               1|                 4|40-50 years|      Average Family|     48694|    0|   3|
|    Village families|               1|                 2|50-60 years|Family with grown...|     49575|    0|   1|
|    Village families|               1|                 3|30-40 years|Family with grown...|     49097|    0|   2|
|    Village families|               1|                 3|40-50 years|Family with grown...|     48108|    0|   3|
|Young all america...|               1|                 3|40-50 years|      Average Family|     48433|    0|   1|
|Young all america...|               2|                 4|40-50 years|      Average Family|     47538|    0|   2|
|Young all america...|               1|                 4|30-40 years|      Average Family|     47393|    0|   3|
|Young urban have-...|               2|                 2|50-60 years|         Living well|     31511|    0|   1|
|Young urban have-...|               1|                 2|30-40 years|         Living well|     26788|    0|   2|
+--------------------+----------------+------------------+-----------+--------------------+----------+-----+----+
only showing top 20 rows
```

Conclusion

In this chapter, I discussed different techniques to read, clean, and preprocess data in PySpark. You saw the methods to join a dataframe and create a pivot table from it. The final sections of the chapter covered UDFs and window-based operations in PySpark. The upcoming chapters will focus on handling streaming data in PySpark and machine learning using MLlib.

CHAPTER 3

Spark Structured Streaming

This chapter discusses how to use Spark's streaming API to process real-time data. The first part focuses on the main difference between streaming and batch data, in addition to their specific applications. The second section provides details on the Structured Streaming API and its various improvements over previous RDD-based Spark streaming APIs. The final section includes the code to use for Structured Streaming on incoming data and discusses how to save the output results in memory. We'll also look at an alternative to Structured Streaming.

Batch vs. Stream

Perhaps most readers of this book are already familiar with the key distinction between batch vs. stream data processing. Nonetheless, we can start on this note, as it emphasizes the importance of stream processing today. If we think of data as a huge ocean, then batch data can be referred to as a bucket of water, and we can have multiple buckets of different sizes, whereas stream data can be considered to be a water pipe that is continuously pumping water from the ocean.

Batch Data

As the name suggests, *batch* refers to a group of records put together over a period of time and later used for processing and analysis. Because these records are collected over a period of time, size-wise, batch data is generally bigger than streaming data (in some cases, however, stream data can be bigger than batch data) and is often used to conduct postmortems for various analysis purposes. The legacy systems, SQL databases, and mainframes all fall under the category of batch data. The key difference compared to streaming data is that batch data is not processed as soon as it becomes part of an earlier batch dataset.

Stream Processing

Stream processing refers to the processing of records in real time or near real time. One doesn't wait for the day to end to then process or analyze the data. Rather, records of the dataset are processed one by one as soon as they become available or based on a window period, as shown in Figure 3-1. As a result, this creates a sort of infinite table, with records continuously being added as data flows from the stream.

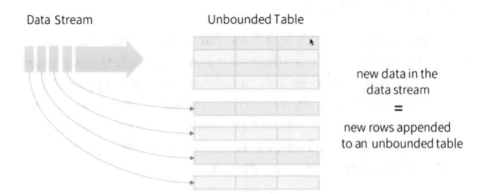

Figure 3-1. *Streaming data*

Today, businesses are very aggressive about using real-time data from various sources, such as platforms, devices, applications, and system logs, in order to keep their competitive edge. Therefore, stream processing has become a critical part of that overall process. Businesses want to use the latest or freshest data, to generate useful insights that can help in decision making. Batch processing cannot offer analytics on the fly, as it doesn't work on a real-time basis, whereas stream data processing can help more effectively in such cases as fraud detection.

Spark Streaming

In the previous chapter, you have seen the core architecture of Spark. One of the components of the Spark framework is Spark Streaming (Structured Streaming), shown in Figure 3-2.

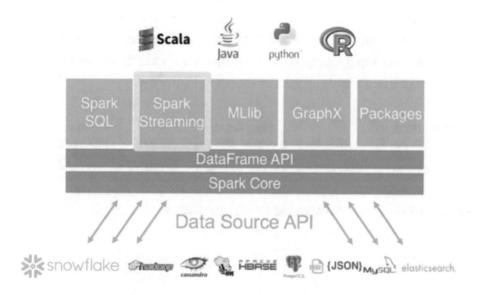

Figure 3-2. Spark Streaming

The earlier version of Spark offered a streaming API that was known as Spark Streaming (Dstream). Spark Streaming was based on RDDs (an earlier Spark abstraction before DataFrame/datasets) and had few limitations. As shown in Figure 3-3, it was able to receive input data from various sources, such as Kafka, Flume, etc., and convert the incoming data into micro-batches and process them using Spark Engine.

Figure 3-3. *Spark Streaming data flow*

The results of each batch would be generated as stream-only and would be saved to the output location. Each micro-batch was an RDD, based on certain time intervals, as shown in Figure 3-4.

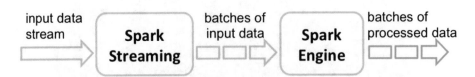

Figure 3-4. *Spark Streaming batch processing*

Although the earlier Spark streaming component was quite powerful in terms of handling the streaming data processing, it was lacking in certain aspects.

1. One core drawback was that there were different APIs for batch and stream data processing jobs. Lots of changes (translations) had to be made to convert a batch job into a Dstream job.

2. It was unable to handle batch data processing based on event time, only on batch time. It was difficult to manage late arriving data for processing.

3. It had limited fault tolerance capability, without any end-to-end guarantee of consistent data processing.

Structured Streaming

The latest version of the streaming component in Spark is known as Structured Streaming, which is a huge improvement over the last RDD-based Spark streaming API. The first significant change from the previous version is that Structured Streaming offers the same API for batch as well as stream data processing jobs. Therefore, it works in a similar way for static and bounded batch data as for streaming and unbounded data, as shown in Figure 3-5.

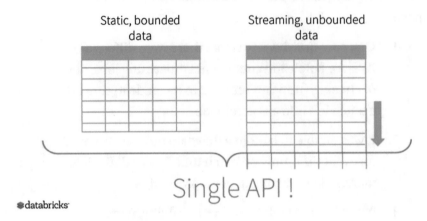

Figure 3-5. *Static vs. Streaming DataFrame*

Another major shift from earlier versions is that Spark Structured Streaming is now built on top of the Spark SQL engine and uses DataFrame for multiple operations, such as aggregation, filters, etc. It also provides the end-to-end guarantee of data consistency, while writing the results in an output location. In order to understand how Structured Streaming works, let's go through how data flows through its programming model, as shown in Figure 3-6.

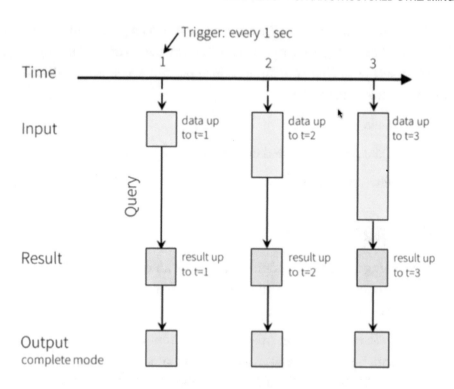

Figure 3-6. *Structured Streaming process*

As the data arrives at time interval 1 (based on the window period selected), the input DataFrame consists of all the records up to that time interval (t=1). The next step is query execution (processing, transformation, join, aggregation) on that particular DataFrame (t=1). Once the query is completed, the results are made available, to be saved in the relevant output (console, memory, location). Now, new data arrives at time interval 2 and is added to the earlier DataFrame (t=1), resulting in a larger DataFrame (t=2). The query is again executed, but this time on a new DataFrame (t=2), and the results are saved in the selected output mode. This process continues for incoming streams of records, and each record is appended to the input DataFrame for data processing.

Now that we understand the basic process of data processing with Structured Streaming, we can consider the core pieces of a streaming-based application. There are three main areas into which we can divide this streaming framework:

1. Data input

2. Data processing (real time or near real time)

3. Final output

Data Input

Any streaming application requires data, to be able to ingest and process data continuously. Therefore, there are multiple ways to provide data as input to the Structured Streaming platform.

- *Messaging systems*: Apache Kafka, Flume, and Logstash can be used to ingest real-time data and, hence, can easily become part of building streaming pipelines. The idea of using these tools is to capture all the data points as data is generated at the source application (web app, mobile app, IoT device) and pass it along to the Structured Streaming platform for further processing and analysis in a fault-tolerant and scalable manner.

- *File folders/directory*: Files are read continuously from the directory as a stream of data. A scheduler can be used to put the new files into the directory. The files can be in text, Parquet, or JSON format. The only condition is to have all the files available in the same format.

Data Processing

This is at the core of using streaming data for creating business value, as it can be applied to certain operations on the incoming data to get results. Such operations as aggregations, filtering, joins, sorting, etc., can be applied.

Final Output

Structured Streaming provides multiple options for users to save their output results, as required, and it can either be in *Append* or *Complete* mode. Append mode refers to adding only new results to the final output table, whereas Complete mode updates the entire results table at the final output location.

1. File directory sink

2. Console

3. Memory sink

Building a Structured App

In this final section of the chapter, we now build a Structured Streaming app that can read files from the local system folder as new files are added to the folder as stream data and apply all the operations on the new data and, finally, write the results in an output directory. The first step is to create the SparkSession object, in order to use Spark.

```
[In]: from pyspark.sql import SparkSession
[In]: spark=SparkSession.builder.appName('structured_
streaming').getOrCreate()
```

```
[In]: import pyspark.sql.functions as F
[In]: from pyspark.sql.types import *
```

Then, we create some self-generated data that can be pushed into a respective local directory ("csv folder"), to be read by the Structured Streaming. The data that we will generate contains four columns and is in CSV format. We can also generate a Parquet format, if required.

1. User ID

2. App

3. Time spent (secs)

4. Age

```
[In]:df_1=spark.createDataFrame([("XN203",'FB',300,30),
("XN201",'Twitter',10,19),("XN202",'Insta',500,45)],
["user_id","app","time_in_secs","age"]).write.csv
("csv_folder",mode='append')
```

Once we have created these dataframes, we can define the schema of these files, in order to read them using stream processing.

```
[In]:schema=StructType().add("user_id","string").
add("app","string").add("time_in_secs", "integer").add("age",
"integer")
```

Now that we have one file available in the local folder ("csv folder"), we can go ahead and read it as a stream dataframe. The API to read a static dataframe is similar to that for reading a streaming dataframe, the only difference being that we use readStream.

```
[In]: data=spark.readStream.option("sep", ",").schema(schema).
csv("csv_folder")
```

To validate the schema of the dataframe, we can use the printSchema command.

```
[In]: data.printSchema()
[Out]:
```

```
root
 |-- user_id: string (nullable = true)
 |-- app: string (nullable = true)
 |-- time_in_secs: integer (nullable = true)
 |-- age: integer (nullable = true)
```

Operations

Once we have the streaming dataframe available, we can apply multiple transformations, in order to get different results, based on specific requirements. In this example, we are going to see aggregations, sorting, filters, etc. First, is simply to count the records of each app in the dataframe. We can write the command as if we are applying the transformations on the static dataframe.

```
[In]: app_count=data.groupBy('app').count()
```

In order to view the results, we must mention the output mode, in addition to the desired location. In this example, we write the results in memory, but it can be written to console, specific cloud storage, or any other location. We also give the output mode as complete, in order to write results on the entire dataframe every time. Finally, we use a simple Spark SQL command to view the output from the query we executed on the streaming dataframe, by converting to a Pandas dataframe.

```
[In]:query=(app_count.writeStream.queryName('count_query').
outputMode('complete').format('memory').start())
[In]: spark.sql("select * from count_query ").toPandas().head(5)
[Out]:
```

	app	count
0	Insta	1
1	FB	1
2	Twitter	1

In this example, a query is being written to filter only the records of the Facebook (FB) app. The average time spent by each user on the FB app is then calculated.

```
[In]: fb_data=data.filter(data['app']=='FB')
[In]: fb_avg_time=fb_data.groupBy('user_id').agg(F.avg("time_
in_secs"))
[In]:fb_query=(fb_avg_time.writeStream.queryName('fb_query').
outputMode('complete').format('memory').start())
[In]: spark.sql("select * from fb_query ").toPandas().head(5)
[Out]:
```

	user_id	avg(time_in_secs)
0	XN203	300.0

Because there is only one dataframe currently in the local folder, we get the output of one user accessing FB and the time spent. In order to view more relative results, let's push more self-generated data to the folder.

```
[In]:df_2=spark.createDataFrame([("XN203",'FB',100,30),("XN201",
'FB',10,19),("XN202",'FB',2000,45)],["user_id","app","time_in_
secs","age"]).write.csv("csv_folder",mode='append')
```

We can now safely assume that Spark Structured Streaming has read the new records and appended them into the streaming dataframe and, therefore, the new results for the same query will differ from the last one.

```
[In]: spark.sql("select * from fb_query ").toPandas().head(5)
```

[Out]:

	user_id	avg(time_in_secs)
0	XN203	200.0
1	XN201	10.0
2	XN202	2000.0

Now, we have the average time spent across all users using the FB app. Let's add few more records to the folder.

```
[In]:df_3=spark.createDataFrame([("XN203",'FB',500,30),
("XN201",'Insta',30,19),("XN202",'Twitter',100,45)],
["user_id","app","time_in_secs","age"]).write.csv("csv_folder",
mode='append')
```

```
[In]: spark.sql("select * from fb_query ").toPandas().head(5)
[Out]:
```

	user_id	avg(time_in_secs)
0	XN203	300.0
1	XN201	10.0
2	XN202	2000.0

In this example, we see aggregation and sorting of the query on the existing dataframe in the local folder. We group all the records by app and calculate the total time spent on each app, in decreasing order.

```
[In]:app_df=data.groupBy('app').agg(F.sum('time_in_secs').
alias('total_time')).orderBy('total_time',ascending=False)
[In]:app_query=(app_df.writeStream.queryName('app_wise_query').
outputMode('complete').format('memory').start())
```

```
[In]: spark.sql("select * from app_wise_query ").toPandas().head(5)
```

[Out]:

	app	total_time
0	FB	2910
1	Insta	530
2	Twitter	110

We now have the results for each app and the total time spent by all users on the respective app, using a stream dataframe. Let's add new records one more time and see the revised results for the same query.

```
[In]:df_4=spark.createDataFrame([("XN203",'FB',500,30),
("XN201",'Insta',30,19),("XN202",'Twitter',100,45)],
["user_id","app","time_in_secs","age"]).write.csv("csv_folder",
mode='append')
```

```
[In]: spark.sql("select * from app_wise_query ").toPandas().
head(5)
[Out]:
```

	app	total_time
0	FB	3410
1	Insta	560
2	Twitter	210

In this example, we try to find the average age of users for every app in our data. We simply group the data by app, take the average age of all the users, and sort the results in decreasing order.

```
[In]:age_df=data.groupBy('app').agg(F.avg('age').alias('mean_
age')).orderBy('mean_age',ascending=False)
```

```
[In]:age_query=(age_df.writeStream.queryName('age_query').
outputMode('complete').format('memory').start())
```

```
[In]:df_5=spark.createDataFrame([("XN210",'FB',500,50),
("XN255",'Insta',30,23),("XN222",'Twitter',100,30)],
["user_id","app","time_in_secs","age"]).write.csv("csv_folder",
mode='append')
```

```
[In]: spark.sql("select * from age_query ").toPandas().head(5)
```

`[Out]:`

	app	mean_age
0	Twitter	36.333333
1	FB	30.666667
2	Insta	27.666667

So, in the preceding examples, we see how we can use Spark Structured Streaming to read the incoming data and create a streaming dataframe to apply various transformations and write the results in a particular location. One more common requirement on streaming data is joins.

Joins

Sometimes we have to merge incoming data with batch data, to make it more comprehensive. In the following example, we will see how we can merge incoming data (stream dataframe) with a static dataframe that contains the full name of the apps. Let's create a new static dataframe with two columns (app and full name).

```
[In]:app_df=spark.createDataFrame([('FB','FACEBOOK'),('Insta',
'INSTAGRAM'),('Twitter','TWITTER')],["app", "full_name"])
[In]: app_df.show()
```

[Out]:

```
+-------+---------+
|    app|full_name|
+-------+---------+
|     FB| FACEBOOK|
|  Insta|INSTAGRAM|
|Twitter|  TWITTER|
+-------+---------+
```

Now that we have a static dataframe available, we can simply write a new query to join the streaming dataframe (data) that we have been working with so far and merge both of them in an app column.

```
[In]: app_stream_df=data.join(app_df,'app')
[In]:join_query=(app_stream_df.writeStream.queryName('join_
query').outputMode('append').format('memory').start())
[In]: spark.sql("select * from join_query ").toPandas().
head(50)
```

[Out]:

	app	user_id	time_in_secs	age	full_name
0	FB	XN201	10	19	FACEBOOK
1	FB	XN203	100	30	FACEBOOK
2	FB	XN203	300	30	FACEBOOK
3	FB	XN202	2000	45	FACEBOOK
4	Insta	XN202	500	45	INSTAGRAM
5	Twitter	XN201	10	19	TWITTER
6	FB	XN203	500	30	FACEBOOK
7	Insta	XN201	30	19	INSTAGRAM
8	Twitter	XN202	100	45	TWITTER
9	FB	XN203	500	30	FACEBOOK
10	Insta	XN201	30	19	INSTAGRAM
11	Twitter	XN202	100	45	TWITTER

As you can see, we now have the additional column (full_name) in the streaming dataframe as well.

Structured Streaming Alternatives

Since Spark Structured Streaming was made available in 2016, it has been rapidly gaining attention from the developers community. Having said that, however, there are a couple of other powerful alternatives to Spark's Structured Streaming. One of them is Flink, which offers similar capability, in terms of streaming data processing with excellent latency rate. Another alternative is Google's Beam, which is suitable in limited cases.

The core advantage of Spark's Structured Streaming over the alternatives is the full-fledged framework provided by Spark as a whole, including batch processing (no major difference in code for batch or stream) and the machine learning library. Another great attribute of Structured Streaming is its Spark SQL API, which is extremely comfortable to many users. It is hoped that new versions of Spark will include more features for Structured Streaming, such as stream joins. Selection depends on the specific requirements of the application and finding the best possible alternative to make it scalable, fault-tolerant, and robust.

Conclusion

In this chapter, the basic difference between batch and streaming data was identified. How Spark's streaming API has evolved over the last few years to become the default framework for building streaming data applications was then discussed. Examples of how Spark's Structured Streaming API is used to read streaming data (local folders), and how to save the aggregated results, were then provided.

CHAPTER 4

Airflow

This chapter focuses on introducing Airflow and how it can be used to handle complex data workflows. Airflow was developed in-house by Airbnb engineers, to manage internal workflows in an efficient manner. Airflow later went on to become part of Apache in 2016 and was made available to users as an open source. Basically, Airflow is a framework for executing, scheduling, distributing, and monitoring various jobs in which there can be multiple tasks that are either interdependent or independent of one another. Every job that is run using Airflow must be defined via a directed acyclic graph (DAG) definition file, which contains a collection you want to run, grouped by relationships and dependencies.

This chapter concentrates on three main topics. I'll start by examining workflow and then cover the basic building block of Airflow: DAG. You will then learn about the user interface aspect of Airflow. In the final section, I will go over the code to define DAG for a job and how to use Airflow to execute and monitor it.

Workflows

Most of the things that we see around us follow a process. Trains run at set intervals, planes fly on fixed times, and signals on roads change at regular periods. Process is critical if consistency is required, especially when there are dependencies among different tasks. In the world of software or technology, a set process is also followed, in building or executing projects. If we go down a few levels, we can call these processes workflows. A proper

© Pramod Singh 2019
P. Singh, *Learn PySpark*, https://doi.org/10.1007/978-1-4842-4961-1_4

workflow must be designed and executed, in order to achieve the desired results. It is rare to witness developers or engineers who don't follow a standard procedure, to build a solution or application. For example, in a typical scenario for data processing workflow, developers will define the steps before executing the tasks, as shown in Figure 4-1. These steps are often referred to as pipelines, which consist of sequences of tasks that must be completed.

Figure 4-1. *Sample workflow*

The first task is to read the data from the source file and then copy it to your platform or desired location. Once data is made available, you ingest it and pass it through multiple steps to clean and transform it. You then perform the analysis and calculate the results. Finally, you save the output to the required location. Traditionally, the way these steps are executed is with the help of cron jobs (commands run by the cron daemon at scheduled intervals). So, all the tasks are part of the scripts which are run as cron jobs. Although, it serves the purpose of executing things in a particular order, it still faces many practical challenges. Most common is the breakdown of the script execution. There is no clear methodology to re-attempt the script run. Another challenge is monitoring the status of the running job. It is very difficult to determine the stage and the time duration each stage takes while executing. There are other challenges as well, such as running multiple cron jobs through a centralized scheduler, for bigger and complex pipelines and to accommodate continuous changes in the workflow. Owing to all of the difficulties mentioned, Airflow was born as a framework with which to schedule and monitor such kinds of jobs and run the pipelines smoothly, for consistent output. To truly understand how Airflow and DAGs operate, you first must understand graphs in general.

Graph Overview

A typical graph data structure consists of two entities, shown in Figure 4-2.

1. Edges

2. Nodes/vertices

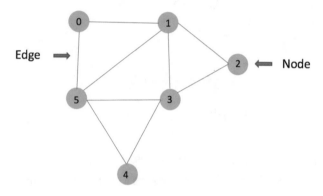

Figure 4-2. *Structure of a graph*

The edges are essentially the connections between the nodes/vertices, and nodes are where actual data resides. We can place graph-based networks into two main categories:

1. Undirected graphs

2. Directed graphs

Undirected Graphs

In this kind of graph structure, the edges or connections don't have any direction, as in Figure 4-2. The relationship will exist at both ends. For example, if person 1 in node 1 is a friend of person 2 in node 2, then person 2 would also be a friend of person 1.

Directed Graphs

A directed graph is either cyclic or acyclic. The direction of the edge plays an important role, as all the edges in a graph go only one way, as in Figure 4-3.

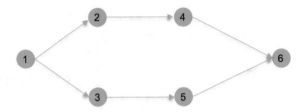

Figure 4-3. *Directed graph*

Cyclic graphs have one or more cycles. A cycle is a path that begins and ends at the same node, as shown in Figure 4-4. The information flows from node 1 to node 2, but there is another way back from node 2 to node 1. This is known as a cycle, or loop, graph.

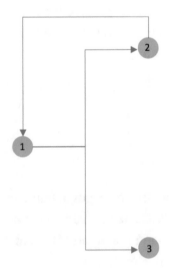

Figure 4-4. *Cyclic graph*

There are no cycles in DAGs, so their benefits include the following:

- A dynamic framework (configuration as code)

- Extensibility—They support different types of task execution.

- Scalability—They can perform an infinite number of tasks (worker nodes).

Let's now turn our attention to DAGs.

DAG Overview

Because a DAG is a directed graph, information can flow in only one direction, and that's forward, as illustrated in Figure 4-5. So, if node 4 must be reached, the path is 1 ➤ 2 ➤ 4. In DAGs, there is no reverse path back to the starting node.

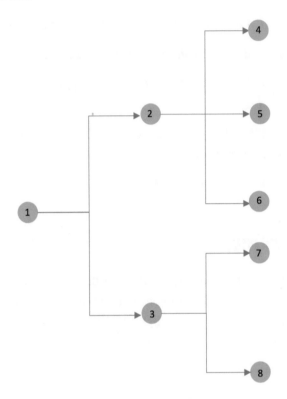

Figure 4-5. *Flow of information in a DAG*

All the tasks of jobs in Airflow must be defined in a DAG. So, the order of execution is defined in DAG form. For example, if task 8 has to be executed, task 1 and task 3 must be finished first, whereas for task 7 to be completed, there doesn't have to be any dependence on other tasks, apart from task 3. Therefore, some of the independent tasks (2, 3) can take place irrespective of each other's state as shown in Figure 4-5. The order of execution of the tasks and interdependencies can be defined well before executing them with Airflow.

All the configurations related to DAG are defined in a DAG definition file, which is a Python extension. It contains all the dependencies and configuration parameters, such as e-mail to be sent in case of failures, start time, end time, and number of retries. We also have to define all the tasks that are part of the DAG, in addition to the dependencies or sequence of the tasks.

Operators

As discussed previously, a DAG can contain multiple tasks. These tasks can be totally different from one another. One of the tasks can be a simple Python script; another can be a shell script or SQL query; and another can be a cloud-based Spark job. These tasks are defined inside of DAG definition file, using operators. Airflow provides a range of operators for different types of tasks. The most common ones are the following:

1. Python operator (Python script)

2. Bash operator (Shell script)

3. SQL operator

4. Docker operator

5. Cloud operator (S3, Azure, Google)

This flexibility of Airflow to run any type of task makes it very powerful, compared to other schedulers.

A DAG file is simply a Python script that contains all the required configuration parameters to run the DAG. There are a few standard steps that must be taken, in order to create a DAG, as shown following:

1. Importing required modules

2. Declaring default arguments

3. Instantiating the DAG object

4. Defining all the tasks

5. Declaring the order of execution/task dependencies

In some areas, Airflow stands out, compared to alternatives, in terms of handling upstream and downstream dependencies more intelligently and in its ability to run historical loads. It also handles failures and blockers with complete transparency.

Installing Airflow

There are multiple ways in which Airflow can be used, as it is designed to easily integrate with different environments. If we want to use Airflow locally, we can install and configure the Airflow environment locally. The other option is to make use of Docker, to containerize the Airflow application and run it on any platform, irrespective of the environment. The core benefit of using Docker is that it takes on itself the additional burden of managing dependencies and deployment.

Airflow Using Docker

One way to run Airflow is to create your own Docker image, with all the dependencies and components, but an Airflow Docker image that makes it very easy to run is already available from the Docker Hub. The steps to run Airflow are as follows:

```
[In]: docker pull puckel/docker-airflow
[In]:docker run -d -p 8080:8080 puckel/docker-airflow webserver
```

Once you run the preceding command in a terminal, you can access port 8080, by going to http://127.0.0.1:8080, as shown in Figure 4-6.

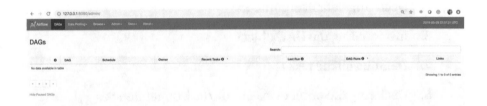

Figure 4-6. Airflow UI

This is how the Airflow UI looks. It currently holds no DAGs and is a vanilla version. If you want to see the sample DAGs that come in Airflow by default, you simply have to add one extra argument while running Docker. To see which containers are running, we can use the following code:

```
[In]: docker ps
```

Once all the containers are listed, you can stop/kill the earlier container running Airflow, by using

```
[In]: docker kill <containerID>
```

Now we run the following command with LOAD_EX=y as an additional parameter, as follows:

```
[In]: docker run -d -p 8080:8080 -e LOAD_EX=y puckel/docker-
airflow
```

If we access the Airflow UI now, we get a list of all default DAGs, as shown in Figure 4-7.

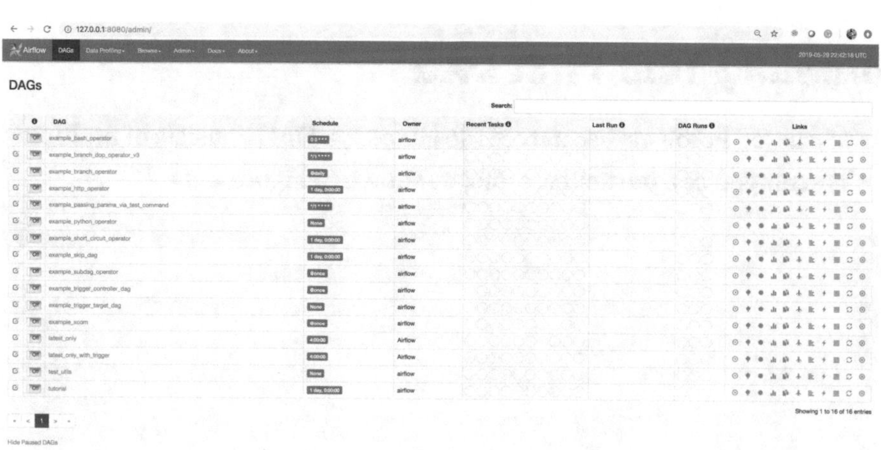

Figure 4-7. *Airflow DAGs*

Airflow Setup (Mac)

The first step is to ensure that Python is installed on the machine, and we can use the brew install command to install python3. Once Python is installed, we can install Airflow. It requires a home directory. ~/airflow is the default, but a different location can also be chosen, based on preference.

```
[In]:  brew install python python3
[In]: pip install airflow
[In]: mkdir ~/airflow
[In]: export AIRFLOW_HOME=~/airflow
 [In]: cd ~/airflow
[In]: airflow initdb
[In]: airflow webserver -p 8080
```

The final two steps are to initialize Airflow, using initdb, and accessing the UI on the preferred port.

Creating Your First DAG

As discussed in the earlier part of the chapter, a DAG consists of multiple tasks arranged in a particular order, as shown in Figure 4-8.

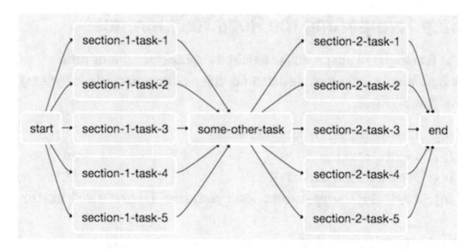

Figure 4-8. *Tasks*

In order to create a DAG, you must define a DAG file that contains all the details pertaining to DAG tasks, and dependencies must be defined in a file (Python script). This is a configuration file specifying the DAG's structure as code. The five steps that must to be taken to run a DAG are shown in Figure 4-9.

Figure 4-9. *Airflow steps*

Let's go into detail about each of these steps, to understand the internals better.

Step 1: Importing the Required Libraries

The first step is to import all the required libraries for running Airflow. Some common ones include datetime, different operators (Bash/Python), and Airflow itself.

```
[In]: from datetime import timedelta
[In]: import airflow
[In]: from airflow import DAG
[In]: from airflow.operators.bash_operator import BashOperator
```

Step 2: Defining the Default Arguments

The next step is to define some important parameters, to ensure that Airflow executes the DAGs at designated time intervals and an appropriate number of times.

```
[In]: args = {
    'owner': 'Pramod',
    'start_date': airflow.utils.dates.days_ago(3),
    # 'end_date': datetime(2018, 12, 30),
    'depends_on_past': False,
    'email': ['airflow@example.com'],
    'email_on_failure': False,
    'email_on_retry': False,
    'retries': 1,
    'retry_delay': timedelta(minutes=5),
    }
```

Step 3: Creating a DAG

The third step is to create the DAG itself, which consists of the DAG's name and schedule interval, as shown following. You can decide when to run the jobs, depending on your requirements.

```
[In]: dag = DAG(
    'pramod_airflow_dag',
    default_args=args,
    description='A simple DAG',
    # Continue to run DAG once per day
    schedule_interval=timedelta(days=1)
```

Step 4: Declaring Tasks

The next step is to declare the tasks (actual jobs) to be executed. All the tasks can be declared and made part of the same DAG created in the preceding step.

```
[In]: t1 = BashOperator(
    task_id='print_date',
    bash_command='date',
    dag=dag,
)

t2 = BashOperator(
    task_id='sleep',
    depends_on_past=False,
    bash_command='sleep 5',
    dag=dag,
)
```

Step 5: Mentioning Dependencies

The final step is to set the order of task execution. They can be either parallel or sequential tasks. There are multiple ways in which the tasks can be defined.

```
[In]: t1 >> t2
```

Once all the preceding steps have been completed, we can start Airflow and access the web UI. The screen shown in Figure 4-10 is available with the DAG file just created.

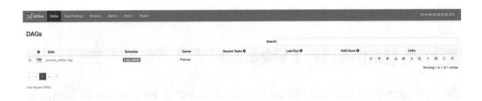

Figure 4-10. *Airflow DAG*

Currently, it's in OFF stage, and we can either trigger it manually or through a command-line interface (CLI). If we click the DAG itself, it takes us to the default tree view of the DAG, which lists all the tasks within the DAG, as shown in Figure 4-11.

Figure 4-11. *Tree view*

Now switch on the DAG, in order to initiate the tasks, as shown in Figure 4-12.

Figure 4-12.   *DAG initialization*

Then click the Trigger Dag icon and start the execution, as shown in Figure 4-13.

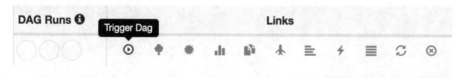

Figure 4-13.   *DAG trigger*

The moment DAG is triggered, we can see the change under the Recent Tasks tab, and DAG runs, as shown in Figure 4-14.

Figure 4-14.   *Running the DAG*

Once the DAG starts to run we can view it in different ways , the first form is the Tree View as shown in Figure 4-15.

Figure 4-15.   *Airflow UI*

Another view is the Graph View that shows the order of task execution in a slightly different manner, and the color of the task indicates its progress (success/running/failed, etc.), as shown in Figure 4-16.

Figure 4-16.   *Graph view*

We can also drill deeper into other details of the DAG, by going to the DAG details tab, shown in Figure 4-17.

Figure 4-17.   *DAG details tab*

Finally, under the Code tab, we can review the DAG Python script that was originally created before running the DAG, as shown in Figure 4-18.

Figure 4-18. *DAG code*

If we wait a while and recheck Graph View, we will observe that both tasks have been completed successfully, as illustrated in Figure 4-19. From the Tree View tab, we can see that tasks have been completed for all three days (the last two and today's), as shown in Figure 4-20.

Figure 4-19. *Graph view*

Figure 4-20. *Tree view*

Conclusion

In this chapter, you learned how to define a DAG and run different jobs using Airflow. The steps to implement Airflow and monitor the status of different tasks were also covered. You were also introduced to the web UI for Airflow, in addition to different components within the interface.

CHAPTER 5

MLlib: Machine Learning Library

Depending on your requirements, there are multiple ways in which you can build machine learning models, using preexisting libraries, such as Python's scikit-learn, R, and TensorFlow. However, what makes Spark's Machine Learning library (MLlib) really useful is its ability to train models on scale and provide distributed training. This allows users to quickly build models on a huge dataset, in addition to preprocessing and preparing workflows with the Spark framework itself.

This chapter focuses on how to leverage MLlib for building and applying various machine learning models. The first part focuses on basic statistics, using MLlib, followed by building pipelines to create features and other transformations. The last part of the chapter discusses using MLlib for building machine learning classification models.

Let's begin by reviewing how we can use Spark's MLlib for calculating some of the basic statistical measures for data analysis. You will see how to calculate correlations between two numerical variables and how to use a chi-square test to determine if there is a significant relationship between two categorical variables.

© Pramod Singh 2019
P. Singh, *Learn PySpark*, https://doi.org/10.1007/978-1-4842-4961-1_5

Calculating Correlations

Correlation is an important metric with which to determine if there is any relationship between two continuous variables. Correlation can either be positive or negative, as shown in Figure 5-1. It is also possible for there to be no correlation between two variables.

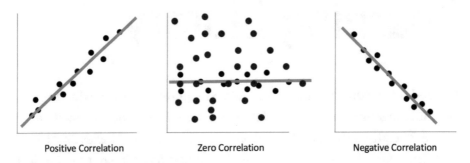

Positive Correlation   Zero Correlation   Negative Correlation

Figure 5-1. *Types of correlations*

Correlation is very easy to calculate using Spark MLlib. It provides the options to calculate two types of coefficients of correlations:

1. Pearson

2. Spearman

Using Spark, let's take a sample dataframe, to calculate the coefficient of correlation. This dataset contains just two numerical columns (Years Experience and Salary). The first step is to create the Spark context, in order to use Spark.

```
[In]: from pyspark.sql import SparkSession
[In]: spark=SparkSession.builder.appName('basic_stats').
getOrCreate()
[In]: df=spark.read.csv('corr_data.csv',header=True,
inferSchema=True)
```

```
[In]: df.count()
[Out]: 30
[In]: df.show()
[Out]:
```

```
+---------------+-------+
|YearsExperience| Salary|
+---------------+-------+
|            1.1|39343.0|
|            1.3|46205.0|
|            1.5|37731.0|
|            2.0|43525.0|
|            2.2|39891.0|
|            2.9|56642.0|
|            3.0|60150.0|
|            3.2|54445.0|
|            3.2|64445.0|
|            3.7|57189.0|
|            3.9|63218.0|
|            4.0|55794.0|
|            4.0|56957.0|
|            4.1|57081.0|
|            4.5|61111.0|
|            4.9|67938.0|
|            5.1|66029.0|
|            5.3|83088.0|
|            5.9|81363.0|
|            6.0|93940.0|
+---------------+-------+
only showing top 20 rows
```

As you can see, there are just 30 records in this dataframe. Next, we combine the two columns into a single dense vector representation, in order to calculate the correlation coefficient, using VectorAssembler. We name the new dense vector "features."

```
[In]: from pyspark.ml.feature import VectorAssembler
[In]: assembler = VectorAssembler(inputCols=df.columns,
outputCol="features")
[In]: df_new=assembler.transform(df)
```

```
[In]: df_new.show()
[Out]:
```

```
+----------------+-------+-------------+
|YearsExperience| Salary|     features|
+----------------+-------+-------------+
|            1.1|39343.0|[1.1,39343.0]|
|            1.3|46205.0|[1.3,46205.0]|
|            1.5|37731.0|[1.5,37731.0]|
|            2.0|43525.0|[2.0,43525.0]|
|            2.2|39891.0|[2.2,39891.0]|
|            2.9|56642.0|[2.9,56642.0]|
|            3.0|60150.0|[3.0,60150.0]|
|            3.2|54445.0|[3.2,54445.0]|
|            3.2|64445.0|[3.2,64445.0]|
|            3.7|57189.0|[3.7,57189.0]|
|            3.9|63218.0|[3.9,63218.0]|
|            4.0|55794.0|[4.0,55794.0]|
|            4.0|56957.0|[4.0,56957.0]|
|            4.1|57081.0|[4.1,57081.0]|
|            4.5|61111.0|[4.5,61111.0]|
|            4.9|67938.0|[4.9,67938.0]|
|            5.1|66029.0|[5.1,66029.0]|
|            5.3|83088.0|[5.3,83088.0]|
|            5.9|81363.0|[5.9,81363.0]|
|            6.0|93940.0|[6.0,93940.0]|
+----------------+-------+-------------+
only showing top 20 rows
```

Pearson Coefficient of Correlation

```
[In]: from pyspark.ml.stat import Correlation
[In]: pearson_corr = Correlation.corr(df_new,'features')
[in]: pearson_corr.show(2,False)
[Out]:
```

```
+--------------------------------------------
|pearson(features)
+--------------------------------------------
|1.0                 0.9782416184887596
0.9782416184887596  1.0                     |
+--------------------------------------------
```

Spearman Coefficient of Correlation

```
[In]: spearman_corr=Correlation.corr(df_
new,'features',"spearman")
[In]: spearman_corr.show(2,False)
[Out]:
```

```
+------------------------------------------.
|spearman(features)
+------------------------------------------.
|1.0                 0.9568313543517044
0.9568313543517044   1.0                 |
+------------------------------------------.
```

Chi-Square Test

Correlation is all about the relationship between numerical features, whereas other types of variables can be categorical as well. One of the ways to validate the relationship between two categorical variables is through a chi-square test. Let's consider an example, to understand how it works. Some made up data is summarized in Table 5-1.

Table 5-1. *Sample Data*

	Smoker	Nonsmoker
Teen	32	12
Young	14	22
Old	6	9

It contains three categories of people (Teen, Young, and Old) and is divided into two buckets (Smoker and Nonsmoker). In the next step, we calculate the total number of people in each category and bucket, as shown in Table 5-2. This is also known as a contingency table.

Table 5-2. *Contingency Table*

	Smoker	Nonsmoker	Total
Teen	32	12	44
Young	14	22	36
Old	6	9	15
	52	43	95

The next step is to calculate the expected values, based on the actual values in the table. The expected values are calculated with the following formula:

$$Expected\ Value_{Teen-Smoker} = (Total_{Teen} * Total_{Smoker})/Total$$
$$Expected\ Value_{Teen-Smoker} = (44 * 52)/95$$
$$Expected\ Value_{Teen-Smoker} = 24$$
$$x^2$$

Similarly, we calculate all the expected values for each category against both buckets, as shown in Table 5-3.

Table 5-3. *Expected Values*

	Smoker	Nonsmoker
Teen	24	20
Young	20	16
Old	8	7

The next step is to arrive at a chi-square table, by comparing the actual values against expected values. The chi-square table values are calculated with the following formula:

$$Chi\text{-}Square\ Table\ Value_{Teen-Smoker} =$$

$$\left(Actual_{Teen-Smoker} - Expected_{Teen-Smoker} \right)^2 / Expected_{Teen-Smoker}$$

$$Chi\text{-}Square\ Table\ Value_{Teen-Smoker} = \frac{(32-24)^2}{24}$$

$$Chi\text{-}Square\ Table\ Value_{Teen-Smoker} = 2.602$$

Now that we have the chi-square values, we take the total for each bucket (Smoker and Nonsmoker), as shown in Table 5-4.

Table 5-4. *Chi-Square Totals*

	Smoker	Nonsmoker
Teen	2.602	3.146
Young	1.652	1.998
Old	0.595	0.720
Chi-square value	4.849	5.864

The overall chi-square value comes to 10.7 (4.84 + 5.86). We then look up for the value of 10.7 in the chi-square table for the degree of freedom (3-1)*(2-1) = 2 and find the corresponding p value. If the p value is less than 0.05, this indicates a statistically significant relationship between the two variables. Let's try to run a chi-square test using Spark. Here we have a sample dataset that has three columns, but we will try to determine if there is any relationship between the marital and housing columns (both are categorical in nature).

```
[In]: df=spark.read.csv('chi_sq.csv',inferSchema=True,header=True)
[In]: df.count()
[Out]: 9501

[In]: df.show()
[Out]:
```

```
+--------+-------+-----+
| marital|housing|label|
+--------+-------+-----+
| married|     no|    0|
| married|     no|    0|
| married|    yes|    0|
| married|     no|    0|
| married|     no|    0|
| married|     no|    0|
| married|     no|    0|
| married|     no|    0|
|  single|    yes|    0|
|  single|    yes|    0|
| married|     no|    0|
|  single|    yes|    0|
|  single|     no|    0|
|divorced|    yes|    0|
| married|    yes|    0|
| married|    yes|    0|
| married|    yes|    0|
| married|    yes|    0|
| married|    yes|    0|
|  single|     no|    0|
+--------+-------+-----+
only showing top 20 rows
```

```
[In]: from pyspark.ml.feature import StringIndexer
[In]: marital_indexer = StringIndexer(inputCol="marital",
outputCol="marital_num").fit(df)
[In]: df = marital_indexer.transform(df)

[In]: from pyspark.ml.feature import OneHotEncoder
[In]: marital_encoder = OneHotEncoder(inputCol="marital_num",
outputCol="marital_vector")
```

```
[In]: df = marital_encoder.transform(df)

[In]: housing_indexer = StringIndexer(inputCol="housing",
outputCol="housing_num").fit(df)
[In]: df = housing_indexer.transform(df)

[In]: housing_encoder = OneHotEncoder(inputCol="housing_num",
outputCol="housing_vector")
[In]: df = housing_encoder.transform(df)

[In]: df.show()

[Out]:
```

```
+--------+-------+-----+------------+--------------+-----------+--------------+
| marital|housing|label|marital_num|marital_vector|housing_num|housing_vector|
+--------+-------+-----+------------+--------------+-----------+--------------+
| married|     no|    0|         0.0| (3,[0],[1.0])|        1.0| (2,[1],[1.0])|
| married|     no|    0|         0.0| (3,[0],[1.0])|        1.0| (2,[1],[1.0])|
| married|    yes|    0|         0.0| (3,[0],[1.0])|        0.0| (2,[0],[1.0])|
| married|     no|    0|         0.0| (3,[0],[1.0])|        1.0| (2,[1],[1.0])|
| married|     no|    0|         0.0| (3,[0],[1.0])|        1.0| (2,[1],[1.0])|
| married|     no|    0|         0.0| (3,[0],[1.0])|        1.0| (2,[1],[1.0])|
| married|     no|    0|         0.0| (3,[0],[1.0])|        1.0| (2,[1],[1.0])|
| married|     no|    0|         0.0| (3,[0],[1.0])|        1.0| (2,[1],[1.0])|
|  single|    yes|    0|         1.0| (3,[1],[1.0])|        0.0| (2,[0],[1.0])|
|  single|    yes|    0|         1.0| (3,[1],[1.0])|        0.0| (2,[0],[1.0])|
| married|     no|    0|         0.0| (3,[0],[1.0])|        1.0| (2,[1],[1.0])|
|  single|    yes|    0|         1.0| (3,[1],[1.0])|        0.0| (2,[0],[1.0])|
|  single|     no|    0|         1.0| (3,[1],[1.0])|        1.0| (2,[1],[1.0])|
|divorced|    yes|    0|         2.0| (3,[2],[1.0])|        0.0| (2,[0],[1.0])|
| married|    yes|    0|         0.0| (3,[0],[1.0])|        0.0| (2,[0],[1.0])|
| married|    yes|    0|         0.0| (3,[0],[1.0])|        0.0| (2,[0],[1.0])|
| married|    yes|    0|         0.0| (3,[0],[1.0])|        0.0| (2,[0],[1.0])|
| married|    yes|    0|         0.0| (3,[0],[1.0])|        0.0| (2,[0],[1.0])|
| married|    yes|    0|         0.0| (3,[0],[1.0])|        0.0| (2,[0],[1.0])|
|  single|     no|    0|         1.0| (3,[1],[1.0])|        1.0| (2,[1],[1.0])|
+--------+-------+-----+------------+--------------+-----------+--------------+
only showing top 20 rows
```

```
[In]: df_assembler = VectorAssembler(inputCols=['marital_
vector','housing_vector'], outputCol="features")
[In]: df = df_assembler.transform(df)

[In]: df.show()
```

[Out]:

```
+-------+-------+-----+-----------+---------------+-----------+---------------+-------------------+
|marital|housing|label|marital_num|marital_vector |housing_num|housing_vector |            features|
+-------+-------+-----+-----------+---------------+-----------+---------------+-------------------+
|married|     no|    0|        0.0| (3,[0],[1.0])|        1.0| (2,[1],[1.0])|(5,[0,4],[1.0,1.0])|
|married|     no|    0|        0.0| (3,[0],[1.0])|        1.0| (2,[1],[1.0])|(5,[0,4],[1.0,1.0])|
|married|    yes|    0|        0.0| (3,[0],[1.0])|        0.0| (2,[0],[1.0])|(5,[0,3],[1.0,1.0])|
|married|     no|    0|        0.0| (3,[0],[1.0])|        1.0| (2,[1],[1.0])|(5,[0,4],[1.0,1.0])|
+-------+-------+-----+-----------+---------------+-----------+---------------+-------------------+
only showing top 4 rows
```

[In]: chi_sq = ChiSquareTest.test(df, "features", "label").head()

[In]: print("pValues: " + str(chi_sq.pValues))

[Out]:
pValues: [0.0,0.0,0.06036632491,0.0,3.56381590905e-14]

Transformations

In this section, I will go over some of the common transformations required for data preprocessing and feature engineering. These help in preparing data the right way for applying machine learning.

Binarizer

We can convert the numerical/continuous variable into categorical features (0/1) by using Binarizer in MLlib. We must declare the threshold value, in order to convert the numerical feature into a binary feature. Any value above the threshold will be converted into 1, and values below or equal to the threshold will become 0. Let's apply Binarizer on the label column of the transformation sample dataset.

```
[In]: df=spark.read.csv('transformations.csv',header=True,infer
Schema=True)
[In]: df.count()
[Out]: 6366

[In]: df.show()
[Out]:
```

```
+-----+-----+-----+-----+-----+-----+-----+-----+---------+
|col_1|col_2|col_3|col_4|col_5|col_6|col_7|col_8|    label|
+-----+-----+-----+-----+-----+-----+-----+-----+---------+
|    3| 32.0|  9.0|  3.0|    3|   17|    2|    5|0.1111111|
|    3| 27.0| 13.0|  3.0|    1|   14|    3|    4|3.2307692|
|    4| 22.0|  2.5|  0.0|    1|   16|    3|    5|1.3999996|
|    4| 37.0| 16.5|  4.0|    3|   16|    5|    5|0.7272727|
|    5| 27.0|  9.0|  1.0|    1|   14|    3|    4| 4.666666|
|    4| 27.0|  9.0|  0.0|    2|   14|    3|    4| 4.666666|
|    5| 37.0| 23.0|  5.5|    2|   12|    5|    4|0.8521735|
|    5| 37.0| 23.0|  5.5|    2|   12|    2|    3| 1.826086|
|    3| 22.0|  2.5|  0.0|    2|   12|    3|    3|4.7999992|
|    3| 27.0|  6.0|  0.0|    1|   16|    3|    5| 1.333333|
|    2| 27.0|  6.0|  2.0|    1|   16|    3|    5|3.2666645|
|    5| 27.0|  6.0|  2.0|    3|   14|    3|    5| 2.041666|
|    3| 37.0| 16.5|  5.5|    1|   12|    2|    3|0.4848484|
|    5| 27.0|  6.0|  0.0|    2|   14|    3|    2|      2.0|
|    4| 22.0|  6.0|  1.0|    1|   14|    4|    4|3.2666645|
|    4| 37.0|  9.0|  2.0|    2|   14|    3|    6|1.3611107|
|    4| 27.0|  6.0|  1.0|    1|   12|    3|    5|      2.0|
|    1| 37.0| 23.0|  5.5|    4|   14|    5|    2| 1.826086|
|    2| 42.0| 23.0|  2.0|    2|   20|    4|    4| 1.826086|
|    4| 37.0|  6.0|  0.0|    2|   16|    5|    4| 2.041666|
+-----+-----+-----+-----+-----+-----+-----+-----+---------+
only showing top 20 rows
```

Now we import the Binarizer function from the Spark library.

```
[In]: from pyspark.ml.feature import Binarizer
[In]: binarizer = Binarizer(threshold=0.99, inputCol="label",
outputCol="binarized_label")

[In]: new_df=binarizer.transform(df)
```

```
[In]: new_df.show()
[Out]:
```

col_1	col_2	col_3	col_4	col_5	col_6	col_7	col_8	label	binarized_label
3	32.0	9.0	3.0	3	17	2	5	0.1111111	0.0
3	27.0	13.0	3.0	1	14	3	4	3.2307692	1.0
4	22.0	2.5	0.0	1	16	3	5	1.3999996	1.0
4	37.0	16.5	4.0	3	16	5	5	0.7272727	0.0
5	27.0	9.0	1.0	1	14	3	4	4.666666	1.0
4	27.0	9.0	0.0	2	14	3	4	4.666666	1.0
5	37.0	23.0	5.5	2	12	5	4	0.8521735	0.0
5	37.0	23.0	5.5	2	12	2	3	1.826086	1.0
3	22.0	2.5	0.0	2	12	3	3	4.7999992	1.0
3	27.0	6.0	0.0	1	16	3	5	1.333333	1.0
2	27.0	6.0	2.0	1	16	3	5	3.2666645	1.0
5	27.0	6.0	2.0	3	14	3	5	2.041666	1.0
3	37.0	16.5	5.5	1	12	2	3	0.4848484	0.0
5	27.0	6.0	0.0	2	14	3	2	2.0	1.0
4	22.0	6.0	1.0	1	14	4	4	3.2666645	1.0
4	37.0	9.0	2.0	2	14	3	6	1.3611107	1.0
4	27.0	6.0	1.0	1	12	3	5	2.0	1.0
1	37.0	23.0	5.5	4	14	5	2	1.826086	1.0
2	42.0	23.0	2.0	2	20	4	4	1.826086	1.0
4	37.0	6.0	0.0	2	16	5	4	2.041666	1.0

```
only showing top 20 rows
```

Principal Component Analysis

Most of the time, we deal with multidimensional data, and it sometimes becomes difficult to understand the underlying pattern without visualizing that data. Principal component analysis (PCA) is one of the transformation techniques that allows you to reduce the dimensions of the data while keeping intact the variation of the data as much as possible. Let's go over the steps to apply PCA, on the same data used previously.

```
[In]: from pyspark.ml.feature import PCA
[In]: assembler = VectorAssembler(inputCols=[col for col in
df.columns if col !='label'], outputCol="features")
[In]: df_new=assembler.transform(df)
```

```
[In]: df_new.show()
```

```
[Out]:
```

```
+-----+-----+-----+-----+-----+-----+-----+-----+---------+--------------------+
|col_1|col_2|col_3|col_4|col_5|col_6|col_7|col_8|    label|            features|
+-----+-----+-----+-----+-----+-----+-----+-----+---------+--------------------+
|    3| 32.0|  9.0|  3.0|    3|   17|    2|    5|0.1111111|[3.0,32.0,9.0,3.0...|
|    3| 27.0| 13.0|  3.0|    1|   14|    3|    4|3.2307692|[3.0,27.0,13.0,3....|
|    4| 22.0|  2.5|  0.0|    1|   16|    3|    5|1.3999996|[4.0,22.0,2.5,0.0...|
|    4| 37.0| 16.5|  4.0|    3|   16|    5|    5|0.7272727|[4.0,37.0,16.5,4....|
|    5| 27.0|  9.0|  1.0|    1|   14|    3|    4| 4.666666|[5.0,27.0,9.0,1.0...|
|    4| 27.0|  9.0|  0.0|    2|   14|    3|    4| 4.666666|[4.0,27.0,9.0,0.0...|
|    5| 37.0| 23.0|  5.5|    2|   12|    5|    4|0.8521735|[5.0,37.0,23.0,5....|
|    5| 37.0| 23.0|  5.5|    2|   12|    2|    3| 1.826086|[5.0,37.0,23.0,5....|
|    3| 22.0|  2.5|  0.0|    2|   12|    3|    3|4.7999992|[3.0,22.0,2.5,0.0...|
|    3| 27.0|  6.0|  0.0|    1|   16|    3|    5| 1.333333|[3.0,27.0,6.0,0.0...|
|    2| 27.0|  6.0|  2.0|    1|   16|    3|    5|3.2666645|[2.0,27.0,6.0,2.0...|
|    5| 27.0|  6.0|  2.0|    3|   14|    3|    5| 2.041666|[5.0,27.0,6.0,2.0...|
|    3| 37.0| 16.5|  5.5|    1|   12|    2|    3|0.4848484|[3.0,37.0,16.5,5....|
|    5| 27.0|  6.0|  0.0|    2|   14|    3|    2|      2.0|[5.0,27.0,6.0,0.0...|
|    4| 22.0|  6.0|  1.0|    1|   14|    4|    4|3.2666645|[4.0,22.0,6.0,1.0...|
|    4| 37.0|  9.0|  2.0|    2|   14|    3|    6|1.3611107|[4.0,37.0,9.0,2.0...|
|    4| 27.0|  6.0|  1.0|    1|   12|    3|    5|      2.0|[4.0,27.0,6.0,1.0...|
|    1| 37.0| 23.0|  5.5|    4|   14|    5|    2| 1.826086|[1.0,37.0,23.0,5....|
|    2| 42.0| 23.0|  2.0|    2|   20|    4|    4| 1.826086|[2.0,42.0,23.0,2....|
|    4| 37.0|  6.0|  0.0|    2|   16|    5|    4| 2.041666|[4.0,37.0,6.0,0.0...|
+-----+-----+-----+-----+-----+-----+-----+-----+---------+--------------------+
only showing top 20 rows
```

k represents the number of reduced dimensions of the data after PCA.

```
[In]: pca = PCA(k=2, inputCol="features", outputCol="pca_
features")
[In]: pca_model=pca.fit(df_new)
```

```
[In]: pca_comp = pca_model.transform(df_new).select("pca_
features")
[In]: pca_comp.show(truncate=False)
[out]:
```

```
+---------------------------------------+
|pca_features                           |
+---------------------------------------+
|[28.476733115993262,24.713719618510776]|
|[27.99124528890644,18.077039274374451] |
|[16.629146085723793,22.253017972879423]|
|[37.443942271456436,23.459943994428997]|
|[24.83901147054452,20.336909763080445] |
|[24.753506078678384,20.43856264253062] |
|[42.331794965931046,17.43772085934015] |
|[42.29009551647075,16.927990063261166] |
|[16.656955524080463,19.436262690743295]|
|[22.571731119237647,23.263203650361966]|
|[22.80481461926103,22.98949375366442]  |
|[22.81498405096786,21.817881649253277] |
|[37.57866646519737,20.035316320032365] |
|[22.51976450217741,21.70529841065548]  |
|[19.291469281017967,19.153104704061747]|
|[31.78912008760198,25.929368868540426] |
|[22.714003671555744,20.605579377862227]|
|[42.34279188958259,18.389657022454944] |
|[45.273795810874965,25.525559532913423]|
|[29.338885487010824,28.955299006778354]|
+---------------------------------------+
only showing top 20 rows
```

As you can see, we have applied PCA to the data (except for the "label" column) and reduced the number of dimensions to just two.

Normalizer

Normalization refers to transformation of data in such a way that the new normalized data has a mean of 0 and a standard deviation of 1. The normalization is done using the following formula:

$$\frac{(x - mean(x))}{standard\ dev(x)}$$

To make use of Normalizer in Spark, we simply have to apply it on the required column. Here, we apply it to the "features" column.

```
[In]: from pyspark.ml.feature import Normalizer
[In]: normalizer = Normalizer(inputCol="features",
outputCol="norm_features", p=1.0)
[In]: normalised_l1_data = normalizer.transform(df_new)
[In]: normalised_l1_data.select('norm_features').
show(truncate=False)
[Out]:
```

```
+-----------------------------------------------------------------------------------------------------------
--------------------------------------------------+
|norm_features
|
+-----------------------------------------------------------------------------------------------------------
--------------------------------------------------+
|[0.04054054054054054,0.43243243243243246,0.12162162162162163,0.04054054054054054,0.04054054054054054,0.2297297297297
2974,0.02702702702702703,0.06756756756756757]    |
|[0.04411764705882353,0.39705882352941174,0.19117647058823528,0.04411764705882353,0.014705882352941176,0.205882352941
17646,0.04411764705882353,0.05882352941176471]  |
|[0.07476635514018691,0.411214953271028,0.04672897196261682,0.0,0.018691588785046728,0.29906542056074764,0.0560747663
55140186,0.09345794392523364]                    |
|[0.04419889502762431,0.4088397790055249,0.18232044198895028,0.04419889502762431,0.03314917127071823,0.17679558011049
723,0.055248618784530384,0.055248618784530384]   |
|[0.078125,0.421875,0.140625,0.015625,0.015625,0.21875,0.046875,0.0625]
|
|[0.06349206349206349,0.42857142857142855,0.14285714285714285,0.0,0.031746031746031744,0.2222222222222222,0.047619047
619047616,0.06349206349206349]                    |
|[0.05347593582887004,0.39572192513368987,0.24598930481283424,0.05882352941176470,0.0213903743315508,0.128342245989
30483,0.053475935882877004,0.0427807486631016]   |
|[0.05586592178770494,0.4134078212290503,0.2569832402234637,0.061452513966480445,0.0223463687150838,0.13407821229050
28,0.0223463687150838,0.0335195530726257]        |
|[0.06315789473684211,0.4631578947368421,0.05263157894736842,0.0,0.042105263157894736,0.25263157894736843,0.063157894
73684211,0.06315789473684211]                    |
|[0.04918032786885246,0.4426229508196721,0.09836065573770492,0.0,0.01639344262295082,0.26229508196721313,0.0491803278
6885246,0.08196721311475409]                      |
|[0.03225806451612903,0.43548387096774194,0.0967741935483871,0.03225806451612903,0.016129032258064516,0.2580645161290
3225,0.04838709677419355,0.08064516129032258]    |
|[0.07692307692307693,0.4153846153846154,0.09230769230769231,0.03076923076923077,0.046153846153846156,0.2153846153846
154,0.046153846153846156,0.07692307692307693]    |
|[0.0375,0.4625,0.20625,0.06875,0.0125,0.15,0.025,0.0375]
|
```

The normalization helps standardize the input data and sometimes improve the performance of the machine learning models.

Standard Scaling

Scaling is another technique to normalize data, such that the values are within a specific range, e.g., [0, 1]. Many machine learning algorithms are sensitive to the scale of the input data, and, hence, it becomes critical to apply scaling. Scaling can be applied in different ways, but the most fundamental approach is to use the following formula:

$$\frac{x - min(x)}{max(x) - min(x)}$$

```
[In]: from pyspark.ml.feature import StandardScaler
[In]: scaler = StandardScaler(inputCol="features",
                              outputCol="scaled_features",
                              withStd=False, withMean=True)

[In]: scaler_model = scaler.fit(df_new)

[In]: scaled_data = scaler_model.transform(df_new)
[In]: scaled_data.select('scaled_features').
show(truncate=False)
```

[Out]:

```
+----------------------------------------------------+
|scaled_features                                     |
|                                                    |
+----------------------------------------------------+
|[-1.1096449890040905,2.9171379202010783,-0.00942507068805476,1.6031259817782024,0.5738297203895746,2.790135092679861
4,-1.4241281809613606,1.14985862393968]              |
|[-1.1096449890040905,-2.0828620797989217,3.9905749293119452,1.6031259817782024,-1.4261702796104254,-0.20986490732013
863,-0.4241281809613606,0.14985862393967997]         |
|[-0.10964498900409048,-7.082862079798922,-6.509425070688055,-1.3968740182217976,-1.4261702796104254,1.79013509267986
14,-0.4241281809613606,1.14985862393968]             |
|[-0.10964498900409048,7.917137920201078,7.490574929311945,2.603125981778202,0.5738297203895746,1.7901350926798614,1.
5758718190386394,1.14985862393968]                  |
|[0.8903550109959095,-2.0828620797989217,-0.00942507068805476,-0.3968740182217976,-1.4261702796104254,-0.209864907320
13863,-0.4241281809613606,0.14985862393967997]       |
|[-0.10964498900409048,-2.0828620797989217,-0.00942507068805476,-1.3968740182217976,-0.4261702796104254,-0.2098649073
2013863,-0.4241281809613606,0.14985862393967997]|
|[0.8903550109959095,7.917137920201078,13.990574929311945,4.103125981778202,-0.4261702796104254,-2.2098649073201386,
1.5758718190386394,0.14985862393967997]             |
|[0.8903550109959095,7.917137920201078,13.990574929311945,4.103125981778202,-0.4261702796104254,-2.2098649073201386,-
1.4241281809613606,-0.85014137606032]               |
|[-1.1096449890040905,-7.082862079798922,-6.509425070688055,-1.3968740182217976,-0.4261702796104254,-2.20986490732013
86,-0.4241281809613606,-0.85014137606032]           |
|[-1.1096449890040905,-2.0828620797989217,-3.0094250706880548,-1.3968740182217976,-1.4261702796104254,1.7901350926798
614,-0.4241281809613606,1.14985862393968]           |
|[-2.1096449890040905,-2.0828620797989217,-3.0094250706880548,0.6031259817782024,-1.4261702796104254,1.79013509267986
14,-0.4241281809613606,1.14985862393968]            |
|[0.8903550109959095,-2.0828620797989217,-3.0094250706880548,0.6031259817782024,0.5738297203895746,-0.209864907320138
63,-0.4241281809613606,1.14985862393968]            |
|[-1.1096449890040905,7.917137920201078,7.490574929311945,4.103125981778202,-1.4261702796104254,-2.2098649073201386,-
1.4241281809613606,-0.85014137606032]               |
|[0.8903550109959095,-2.0828620797989217,-3.0094250706880548,-1.3968740182217976,-0.4261702796104254,-0.2098649073201
3863,-0.4241281809613606,-1.85014137606032]         |
|[-0.10964498900409048,-7.082862079798922,-3.0094250706880548,-0.3968740182217976,-1.4261702796104254,-0.209864907320
13863,0.5758718190386394,0.14985862393967997]       |
|[-0.10964498900409048,7.917137920201078,-0.00942507068805476,0.6031259817782024,-0.4261702796104254,-0.2098649073201
3863,-0.4241281809613606,2.14985862393968]          |
|[-0.10964498900409048,-2.0828620797989217,-3.0094250706880548,-0.3968740182217976,-1.4261702796104254,-2.20986490732
01386,-0.4241281809613606,1.14985862393968]         |
|[-3.1096449890040905,7.917137920201078,13.990574929311945,4.103125981778202,1.5738297203895746,-0.20986490732013863,
1.575871819038639,-1.85014137606032]                |
|[-2.1096449890040905,12.917137920201078,13.990574929311945,0.6031259817782024,-0.4261702796104254,5.790135092679861,
0.5758718190386394,0.14985862393967997]             |
|[-0.10964498900409048,7.917137920201078,-3.0094250706880548,-1.3968740182217976,-0.4261702796104254,1.79013509267986
14,1.5758718190386394,0.14985862393967997]          |
+----------------------------------------------------+
only showing top 20 rows
```

Min-Max Scaling

Min-max scaling is another version of standard scaling, as it allows you to
rescale the feature values between specific limits (mostly, between 0 and 1).
You can also rescale the values between 0 and 1, using min-max scaling.

```
[In]: from pyspark.ml.feature import MinMaxScaler
[In]: mm_scaler = MinMaxScaler(inputCol="features",
outputCol="mm_scaled_features")

[In]: mm_scaler_model = mm_scaler.fit(df_new)

[In]: rescaled_df = mm_scaler_model.transform(df_new)
```

```
[In]: rescaled_df.select("features", "mm_scaled_features").
show()
[Out]:
```

```
+--------------------+--------------------+
|            features| mm_scaled_features|
+--------------------+--------------------+
|[3.0,32.0,9.0,3.0...|[0.5,0.5918367346...|
|[3.0,27.0,13.0,3....|[0.5,0.3877551020...|
|[4.0,22.0,2.5,0.0...|[0.75,0.183673469...|
|[4.0,37.0,16.5,4....|[0.75,0.795918367...|
|[5.0,27.0,9.0,1.0...|[1.0,0.3877551020...|
|[4.0,27.0,9.0,0.0...|[0.75,0.387755102...|
|[5.0,37.0,23.0,5....|[1.0,0.7959183673...|
|[5.0,37.0,23.0,5....|[1.0,0.7959183673...|
|[3.0,22.0,2.5,0.0...|[0.5,0.1836734693...|
|[3.0,27.0,6.0,0.0...|[0.5,0.3877551020...|
|[2.0,27.0,6.0,2.0...|[0.25,0.387755102...|
|[5.0,27.0,6.0,2.0...|[1.0,0.3877551020...|
|[3.0,37.0,16.5,5....|[0.5,0.7959183673...|
|[5.0,27.0,6.0,0.0...|[1.0,0.3877551020...|
|[4.0,22.0,6.0,1.0...|[0.75,0.183673469...|
|[4.0,37.0,9.0,2.0...|[0.75,0.795918367...|
|[4.0,27.0,6.0,1.0...|[0.75,0.387755102...|
|[1.0,37.0,23.0,5....|[0.0,0.7959183673...|
|[2.0,42.0,23.0,2....|[0.25,1.0,1.0,0.3...|
|[4.0,37.0,6.0,0.0...|[0.75,0.795918367...|
+--------------------+--------------------+
only showing top 20 rows
```

To access the min and max values, we can use the getMin and getMax functions. In order to change the range, we can define the new min and max values by creating a scaler object. Here we rescale the values between -1 and 1.

```
[In]: mm_scaler.getMin()
[Out]: 0.0
```

```
[In]: mm_scaler.getMax()
[Out]: 1.0
```

```
Alter the min max values
```

```
[In]: mm_scaler = MinMaxScaler(inputCol="features",
outputCol="mm_scaled_features", min=-1,max=1)

[In]: mm_scaler_model = mm_scaler.fit(df_new)

[In]: rescaled_df = mm_scaler_model.transform(df_new)

[In]: rescaled_df.select("features", "mm_scaled_features").
show()
[Out]:
```

```
+--------------------+--------------------+
|            features|  mm_scaled_features|
+--------------------+--------------------+
|[3.0,32.0,9.0,3.0...|[0.0,0.1836734693...|
|[3.0,27.0,13.0,3....|[0.0,-0.224489795...|
|[4.0,22.0,2.5,0.0...|[0.5,-0.632653061...|
|[4.0,37.0,16.5,4....|[0.5,0.5918367346...|
|[5.0,27.0,9.0,1.0...|[1.0,-0.224489795...|
|[4.0,27.0,9.0,0.0...|[0.5,-0.224489795...|
|[5.0,37.0,23.0,5....|[1.0,0.5918367346...|
|[5.0,37.0,23.0,5....|[1.0,0.5918367346...|
|[3.0,22.0,2.5,0.0...|[0.0,-0.632653061...|
|[3.0,27.0,6.0,0.0...|[0.0,-0.224489795...|
|[2.0,27.0,6.0,2.0...|[-0.5,-0.22448979...|
|[5.0,27.0,6.0,2.0...|[1.0,-0.224489795...|
|[3.0,37.0,16.5,5....|[0.0,0.5918367346...|
|[5.0,27.0,6.0,0.0...|[1.0,-0.224489795...|
|[4.0,22.0,6.0,1.0...|[0.5,-0.632653061...|
|[4.0,37.0,9.0,2.0...|[0.5,0.5918367346...|
|[4.0,27.0,6.0,1.0...|[0.5,-0.224489795...|
|[1.0,37.0,23.0,5....|[-1.0,0.591836734...|
|[2.0,42.0,23.0,2....|[-0.5,1.0,1.0,-0....|
|[4.0,37.0,6.0,0.0...|[0.5,0.5918367346...|
+--------------------+--------------------+
only showing top 20 rows
```

MaxAbsScaler

MaxAbsScaler is a little different from standard scaling tools, as it rescales each feature value between -1 and 1. However, it does not shift the center of the data and, hence, does not impact any sparsity.

```
[In]: from pyspark.ml.feature import MaxAbsScaler
[In]: mxabs_scaler = MaxAbsScaler(inputCol="features",
outputCol="mxabs_features")

[In]: mxabs_scaler_model = mxabs_scaler.fit(df_new)

[In]: rescaled_df = mxabs_scaler_model.transform(df_new)

[In]: rescaled_df.select("features", "mxabs_features").show()

[Out]:
```

```
+--------------------+--------------------+
|            features|       mxabs_features|
+--------------------+--------------------+
|[3.0,32.0,9.0,3.0...|[0.6,0.7619047619...|
|[3.0,27.0,13.0,3....|[0.6,0.6428571428...|
|[4.0,22.0,2.5,0.0...|[0.8,0.5238095238...|
|[4.0,37.0,16.5,4....|[0.8,0.8809523809...|
|[5.0,27.0,9.0,1.0...|[1.0,0.6428571428...|
|[4.0,27.0,9.0,0.0...|[0.8,0.6428571428...|
|[5.0,37.0,23.0,5....|[1.0,0.8809523809...|
|[5.0,37.0,23.0,5....|[1.0,0.8809523809...|
|[3.0,22.0,2.5,0.0...|[0.6,0.5238095238...|
|[3.0,27.0,6.0,0.0...|[0.6,0.6428571428...|
|[2.0,27.0,6.0,2.0...|[0.4,0.6428571428...|
|[5.0,27.0,6.0,2.0...|[1.0,0.6428571428...|
|[3.0,37.0,16.5,5....|[0.6,0.8809523809...|
|[5.0,27.0,6.0,0.0...|[1.0,0.6428571428...|
|[4.0,22.0,6.0,1.0...|[0.8,0.5238095238...|
|[4.0,37.0,9.0,2.0...|[0.8,0.8809523809...|
|[4.0,27.0,6.0,1.0...|[0.8,0.6428571428...|
|[1.0,37.0,23.0,5....|[0.2,0.8809523809...|
|[2.0,42.0,23.0,2....|[0.4,1.0,1.0,0.36...|
|[4.0,37.0,6.0,0.0...|[0.8,0.8809523809...|
+--------------------+--------------------+
only showing top 20 rows
```

Binning

Binning, or bucketing, is useful in cases in which you want to group continuous features into categories. You can do binning with the help of Bucketizer in Spark. Let's try to bucketize the target (label) column into bins. The splits can be made accordingly, but it's always advisable to start and end the splits with negative infinity to positive infinity, to avoid out of bound errors (especially in cases in which max and min values of the feature are unknown).

```
[In]: from pyspark.ml.feature import Bucketizer
```

```
[In]: df.show(10,False)
[Out]:
```

```
+-----+-----+-----+-----+-----+-----+-----+-----+---------+
|col_1|col_2|col_3|col_4|col_5|col_6|col_7|col_8|label    |
+-----+-----+-----+-----+-----+-----+-----+-----+---------+
|3    |32.0 |9.0  |3.0  |3    |17   |2    |5    |0.1111111|
|3    |27.0 |13.0 |3.0  |1    |14   |3    |4    |3.2307692|
|4    |22.0 |2.5  |0.0  |1    |16   |3    |5    |1.3999996|
|4    |37.0 |16.5 |4.0  |3    |16   |5    |5    |0.7272727|
|5    |27.0 |9.0  |1.0  |1    |14   |3    |4    |4.666666 |
|4    |27.0 |9.0  |0.0  |2    |14   |3    |4    |4.666666 |
|5    |37.0 |23.0 |5.5  |2    |12   |5    |4    |0.8521735|
|5    |37.0 |23.0 |5.5  |2    |12   |2    |3    |1.826086 |
|3    |22.0 |2.5  |0.0  |2    |12   |3    |3    |4.7999992|
|3    |27.0 |6.0  |0.0  |1    |16   |3    |5    |1.333333 |
+-----+-----+-----+-----+-----+-----+-----+-----+---------+
only showing top 10 rows
```

We now define the end points for the splits to occur and create a new column that contains the bins.

```
[In]: splits = [0.0,1.0,2.0,3.0,4.0,5.0,float("inf")]
```

```
[In]: bucketizer = Bucketizer(splits=splits, inputCol="label",
outputCol="label_bins")
```

```
[In]: binned_df = bucketizer.transform(df)
```

```
[In]: binned_df.select(['label','label_bins']).show(10,False)
```

[Out]:

```
+----------+----------+
|label     |label_bins|
+----------+----------+
|0.1111111 |0.0       |
|3.2307692 |3.0       |
|1.3999996 |1.0       |
|0.7272727 |0.0       |
|4.666666  |4.0       |
|4.666666  |4.0       |
|0.8521735 |0.0       |
|1.826086  |1.0       |
|4.7999992 |4.0       |
|1.333333  |1.0       |
+----------+----------+
only showing top 10 rows
```

As you can see, all the values have been put under a bin, and we can use groupby to validate the total bins (6). We can also get the bin values using the getSplit function.

[In]: binned_df.groupBy('label_bins').count().show()
[Out]:

```
+----------+-----+
|label_bins|count|
+----------+-----+
|       0.0| 5247|
|       1.0|  429|
|       4.0|  172|
|       3.0|  130|
|       2.0|  239|
|       5.0|  149|
+----------+-----+
```

[In]: print(bucketizer.getSplits())-1
[Out]: 6

Building a Classification Model

In this section, you will see how we can use Spark's Machine Learning library (MLlib) to build classification models. Because there are dedicated chapters for supervised and unsupervised ML models later in the book, in this section, I will not go too deep into the details but focus instead on the overall process of building a model with the MLlib. For our example, we will use the data inspired by the dataset provided by Giulio Palombo in his book *A Collection of Data Science Take-Home Challenges*. The dataset contains information pertaining to a few customers who have applied for new bank loans and whether they will default. We will build a binary classification model to predict whether a particular customer should be granted a loan, based on the knowledge gleaned from the model. The following core steps are used to build a classification model:

1. Load the dataset.

2. Perform exploratory data analysis.

3. Perform required data transformations.

4. Split data into train and test subsets.

5. Train and evaluate the baseline model on train data.

6. Perform hyperparameter tuning.

7. Build a final model with the best parameters.

Step 1: Load the Dataset

In the first step, we initiate the Spark object, to use Spark and load the dataset to create the Spark dataframe.

```
[In]: from pyspark.sql import SparkSession
[In]: spark=SparkSession.builder.appName('binary_class').
getOrCreate()
```

```
[In]: df=spark.read.csv('classification_data.csv',inferSchema=
True,header=True)
```

Step 2: Explore the Dataframe

In this step, we explore the different aspects of the data and various columns in the dataframe.

```
[In]: print((df.count(),len(df.columns)))
```

```
[Out]: (46751, 12)
```

The dataframe contains 12 columns and more than 46,000 records. We can view all the columns and datatypes, using the printSchema function.

```
[In]: df.printSchema()
```

```
[Out]:
```

```
root
 |-- loan_id: string (nullable = true)
 |-- loan_purpose: string (nullable = true)
 |-- is_first_loan: integer (nullable = true)
 |-- total_credit_card_limit: integer (nullable = true)
 |-- avg_percentage_credit_card_limit_used_last_year: double (nullable = true)
 |-- saving_amount: integer (nullable = true)
 |-- checking_amount: integer (nullable = true)
 |-- is_employed: integer (nullable = true)
 |-- yearly_salary: integer (nullable = true)
 |-- age: integer (nullable = true)
 |-- dependent_number: integer (nullable = true)
 |-- label: integer (nullable = true)
```

We can use the show or display function to view the top few rows of the dataframe.

```
[In]: df.show(5)
[Out]:
```

```
+-------+------------+-------------+---------------+--------------------+------------------------------------------------+---------
----+---------------+-----------+--------------+---+-----------------+-----+
|loan_id|loan_purpose|is_first_loan|total_credit_card_limit|avg_percentage_credit_card_limit_used_last_year|saving_am
ount|checking_amount|is_employed|yearly_salary|age|dependent_number|label|
+-------+------------+-------------+---------------+--------------------+------------------------------------------------+---------
----+---------------+-------------+---------------+---+-----------------+-----+
|    A_1|    personal|            1|                        7900|                                            0.8|
1103|           6393|           1|        16400| 42|              4|    0|
|    A_2|    personal|            0|                        3300|                                           0.29|
2588|            832|           1|        75500| 56|              1|    0|
|    A_3|    personal|            0|                        7600|                                            0.9|
1651|           8868|           1|        59000| 46|              1|    0|
|    A_4|    personal|            1|                        3400|                                           0.38|
1269|           6863|           1|        26000| 55|              8|    0|
|    A_5|   emergency|            0|                        2600|                                           0.89|
1310|           3423|           1|         9700| 41|              4|    1|
+-------+------------+-------------+---------------+--------------------+------------------------------------------------+---------
----+---------------+-------------+---------------+---+-----------------+-----+
only showing top 5 rows
```

We can use groupby to count the number of positive and negative events in the target column (label).

```
[In]: df.groupBy('loan_label).count().show()
[Out]:
```

```
+-----+-----+
|label|count|
+-----+-----+
|    1|16201|
|    0|30550|
+-----+-----+
```

As you can see, more than one-third of all customers have defaulted on their loans. To understand the data better, we continue with exploratory data analysis. In the following results, we can see that people prefer to apply for a loan mainly for property, operations, and personal reasons.

```
[In]: df.groupBy('loan_purpose').count().show()
[Out]:
```

```
+------------+-----+
|loan_purpose|count|
+------------+-----+
|      others| 6763|
|   emergency| 7562|
|    property|11388|
|  operations|10580|
|    personal|10458|
+------------+-----+
```

Customers seem to have applied for property loans more than any other category of loan.

Step 3: Data Transformation

Because all of the variables in the dataframe are numerical, except for the loan purpose, we must convert them into numerical form, using OneHotEncoder.

```
[In]: from pyspark.ml.feature import OneHotEncoder,
StringIndexer, VectorAssembler
```

```
[In]: loan_purpose_indexer = StringIndexer(inputCol="loan_
purpose", outputCol="loan_index").fit(df)
[In]: df = loan_purpose_indexer.transform(df)
[In]: loan_encoder = OneHotEncoder(inputCol="loan_index",
outputCol="loan_purpose_vec")
[In]: df = loan_encoder.transform(df)
```

```
[In]: df.select(['loan_purpose','loan_index','loan_purpose_
vec']).show(3,False)
```

```
+------------+----------+----------------+
|loan_purpose|loan_index|loan_purpose_vec|
+------------+----------+----------------+
|personal    |2.0       |(4,[2],[1.0])   |
|personal    |2.0       |(4,[2],[1.0])   |
|personal    |2.0       |(4,[2],[1.0])   |
+------------+----------+----------------+
only showing top 3 rows
```

Now that we have converted the original loan-purpose feature into vectorized form, we can use VectorAssembler to create a single-feature vector for model training.

```
[In]: from pyspark.ml.feature import VectorAssembler

[In]: df_assembler = VectorAssembler(inputCols=['is_first_loan',
  'total_credit_card_limit',
  'avg_percentage_credit_card_limit_used_last_year',
  'saving_amount',
  'checking_amount',
  'is_employed',
  'yearly_salary',
  'age',
  'dependent_number',
  'loan_purpose_vec'], outputCol="features")
[In]: df = df_assembler.transform(df)

[In]: df.select(['features','label']).show(10,False)

[Out]:
```

```
+------------------------------------------------------------------+-----+
|features                                                          |label|
+------------------------------------------------------------------+-----+
|[1.0,7900.0,0.8,1103.0,6393.0,1.0,16400.0,42.0,4.0,0.0,0.0,1.0,0.0]|0    |
|[0.0,3300.0,0.29,2588.0,832.0,1.0,75500.0,56.0,1.0,0.0,0.0,1.0,0.0]|0    |
|[0.0,7600.0,0.9,1651.0,8868.0,1.0,59000.0,46.0,1.0,0.0,0.0,1.0,0.0]|0    |
|[1.0,3400.0,0.38,1269.0,6863.0,1.0,26000.0,55.0,8.0,0.0,0.0,1.0,0.0]|0   |
|[0.0,2600.0,0.89,1310.0,3423.0,1.0,9700.0,41.0,4.0,0.0,0.0,0.0,1.0]|1    |
|[0.0,7600.0,0.51,1040.0,2406.0,1.0,22900.0,52.0,0.0,0.0,1.0,0.0,0.0]|0   |
|[1.0,6900.0,0.82,2408.0,5556.0,1.0,34800.0,48.0,4.0,0.0,1.0,0.0,0.0]|0   |
|[0.0,5700.0,0.56,1933.0,4139.0,1.0,32500.0,64.0,2.0,0.0,0.0,1.0,0.0]|0   |
|[1.0,3400.0,0.95,3866.0,4131.0,1.0,13300.0,23.0,3.0,0.0,0.0,1.0,0.0]|0   |
|[0.0,2900.0,0.91,88.0,2725.0,1.0,21100.0,52.0,1.0,0.0,0.0,1.0,0.0]|1     |
+------------------------------------------------------------------+-----+
only showing top 10 rows
```

We now create a new dataframe with just two columns: features and label.

```
[In]: model_df=df.select(['features','label'])
```

Step 4: Splitting into Train and Test Data

We now split the overall data into training and test sets randomly, to avoid any bias in the training.

```
[In]: training_df,test_df=model_df.randomSplit([0.75,0.25])
```

Step 5: Model Training

Now that our training and test data are ready, we can go ahead and train a baseline model, such as logistic regression, with default parameters and check its performance on train and test data.

```
[In]: from pyspark.ml.classification import LogisticRegression
```

```
[In]: log_reg=LogisticRegression().fit(training_df)
```

```
[In]: lr_summary=log_reg.summary
```

```
[In]: lr_summary.accuracy
[Out]: 0.8939298586875679
[In]: lr_summary.areaUnderROC
0.9587456481363935
```

```
[In]: print(lr_summary.precisionByLabel)
```

```
[Out]: [0.9233245149911816, 0.8396318618667535]
```

```
[In]: print(lr_summary.recallByLabel)
```

```
[Out]: [0.914054997817547, 0.8556606905710491]
```

```
[In]: predictions = log_reg.transform(test_df)
```

```
[In]: predictions.show(10)
```

```
+--------------------+-----+--------------------+--------------------+----------+
|            features|label|       rawPrediction|         probability|prediction|
+--------------------+-----+--------------------+--------------------+----------+
|(13,[0,1,2,3,4,7]...|    1|[-3.5396403026354...|[0.02820514546354...|       1.0|
|(13,[0,1,2,3,4,7]...|    0|[0.93924901841113...|[0.71894793767811...|       0.0|
|(13,[0,1,2,3,4,7]...|    1|[1.17583325317274...|[0.76419778468159...|       0.0|
|(13,[0,1,2,3,4,7,...|    1|[-4.5769161134357...|[0.01018183389971...|       1.0|
|(13,[0,1,2,3,4,7,...|    1|[-6.5927367735521...|[0.00136840955779...|       1.0|
|(13,[0,1,2,3,4,7,...|    1|[-5.7710428439141...|[0.00310682335686...|       1.0|
|(13,[0,1,2,3,4,7,...|    1|[-5.1739740421116...|[0.00563014492250...|       1.0|
|(13,[0,1,2,3,4,7,...|    1|[-7.5227422440788...|[5.40355800305344...|       1.0|
|(13,[0,1,2,3,4,7,...|    1|[-5.0884786864601...|[0.00612959168096...|       1.0|
|(13,[0,1,2,3,4,7,...|    1|[-3.6787387009965...|[0.02463271413862...|       1.0|
+--------------------+-----+--------------------+--------------------+----------+
only showing top 10 rows
```

```
[In]: model_predictions = log_reg.transform(test_df)
```

```
[In]: model_predictions = log_reg.evaluate(test_df)
```

```
[In]: model_predictions.accuracy
[Out]: 0.8945984906300347
```

```
[In]: model_predictions.areaUnderROC
[Out]: 0.9594316478468224
```

```
[In]: print(model_predictions.recallByLabel)
[Out]: [0.9129581151832461, 0.8608235010835541]
```

```
[In]: print(model_predictions.precisionByLabel)
[Out]: [0.9234741162452006, 0.8431603773584906]
```

Step 6: Hyperparameter Tuning

So, using a baseline model, we are getting almost 89% accuracy on the test data, and a recall rate of 0.86. Now that we have built the baseline model, we can build a more sophisticted model, such as a random forest model, which is an ensemble method that can improve the accuracy of predictions. You will see how we can tune this model, to find the best possible hyper-parameters.

```
[In]: from pyspark.ml.classification import
RandomForestClassifier
```

First, we build a random forest model with default hyper-parameters, then train it on the training data, so that predictions can be made on the test data.

```
[In]: rf = RandomForestClassifier()
[In]: rf_model = rf.fit(training_df)

[In]: model_predictions = rf_model.transform(test_df)
```

Using cross-validation techniques, we now try to come up with the best hyperparameters for this model.

```
[In]: from pyspark.ml.tuning import ParamGridBuilder,
CrossValidator
[In]: from pyspark.ml.evaluation import
BinaryClassificationEvaluator

[In]: evaluator = BinaryClassificationEvaluator()

[In]: rf = RandomForestClassifier()

[In]: paramGrid = (ParamGridBuilder()
            .addGrid(rf.maxDepth, [5,10,20,25,30])
            .addGrid(rf.maxBins, [20,30,40 ])
            .addGrid(rf.numTrees, [5, 20,50])
            .build())
```

We define the parameter grid with all the possible values for different hyperparameters (maxDepth, maxBins, numTrees) and apply cross-validation, to dtermine the best model. We use five-fold cross-validation in this case (four parts for training, and one for testing).

```
[In]: cv = CrossValidator(estimator=rf, estimatorParam
Maps=paramGrid, evaluator=evaluator, numFolds=5)
[In]: cv_model = cv.fit(training_df)
```

We then access the best model parameters and use them on the test dataset, to make predictions.

Step 7: Best Model

```
[In]: best_rf_model = cv_model.bestModel
[In]: model_predictions = best_rf_model.transform(test_df)

[In]:true_pos=model_predictions.filter(model_
predictions['label']==1).filter(model_
predictions['prediction']==1).count()

[In]:actual_pos=model_predictions.filter(model_
predictions['label']==1).count()

[In]:pred_pos=model_predictions.filter(model_
predictions['prediction']==1).count()
[In]: recall_rate=float(true_pos)/(actual_pos)
[In]: print(recall_rate)

[Out]: 0.912426614481409
```

As you can see from the preceding, with the random forest model with best hyperparameters, the recall rate has improved, compared to the baseline method (logistic regression).

Conclusion

In this chapter, some transformation techniques using PySpark and ways to compute summary statistics were reviewed. You saw how to build a machine learning model from scratch and how to tune hyperparameters, to choose the best parameters for a model.

CHAPTER 6

Supervised Machine Learning

Machine learning can be broadly divided into four categories: supervised machine learning and unsupervised machine learning and, to a lesser extent, semi-supervised machine learning and reinforcement machine learning. Because supervised machine learning drives a lot of business applications and significantly affects our day-to-day lives, it is considered one of the most important categories.

This chapter reviews supervised machine learning, using multiple algorithms. In Chapter 7, we'll look at unsupervised machine learning. I'll begin by providing an overview of the different categories of supervised machine learning. In the second section, I will cover various regression methods, and we will build machine learning models, using PySpark's MLlib library. The third and final section of this chapter focuses on classification, using multiple machine learning algorithms.

Supervised Machine Learning Primer

In supervised machine learning, as the name suggests, the learning process is supervised, as the machine learning algorithm being used corrects its predictions, based on the actual output. In supervised machine learning, the correct labels or output is already known during the model

© Pramod Singh 2019
P. Singh, *Learn PySpark*, https://doi.org/10.1007/978-1-4842-4961-1_6

training phase, and, hence, the error can be reduced accordingly. In short, we try to map the relationship between the input data and output label in such a way as to pick up the signals from the training data and generalize about the unseen data as well. The training of the model consists of comparing the actual output with the predicted output and then making the changes in predictions, to reduce the total error between what is actual and what is predicted. The supervised machine learning process followed is as shown in Figure 6-1.

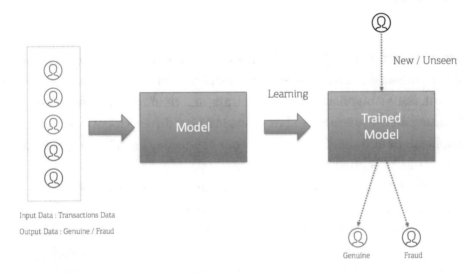

Figure 6-1. *Supervised learning approach*

The data used for training the model is preprocessed, and features are created accordingly. Once the machine learning model is trained, it can be used to make predictions on the unseen data. So, in the preceding figure, we can see how the model is trained, using input data and how now, the trained model is used to predict whether the new transaction is genuine. This type of learning is predominantly used in cases in which historical data is available and predictions must be made on future data. The further categorization of supervised learning is based on types of output or target variables being used for prediction:

- Regression

- Classification

Regression is used when the target value that is being predicted is continuous or numerical in nature. For example, predicting salary based on a given number of years of experience or education falls under the category of regression.

Note   Although there are multiple types of regression, in this chapter, I'll focus on linear regression and some of its associated algorithms, as you'll see shortly.

Classification is used if the target variable is a discrete value or categorical in nature. For example, predicting whether a customer will churn out is a type of classification problem, as shown in Figure 6-2.

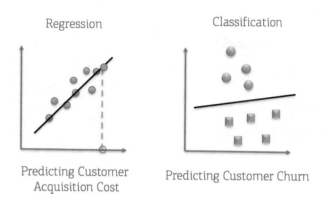

Figure 6-2. *Types of suprvised tasks*

119

Classification tasks can further be broken down into two categories: binary class and multi-class, as shown in Figure 6-3.

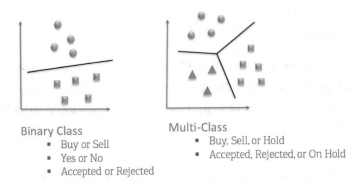

Binary Class
- Buy or Sell
- Yes or No
- Accepted or Rejected

Multi-Class
- Buy, Sell, or Hold
- Accepted, Rejected, or On Hold

Figure 6-3. *Types of classes*

Binary Classification

When the target or output variable contains only up to two categories, it is referred to as binary classification. So, every record in the data can only fall under one of the two groups. For example:

- Yes or no

- Group A or group B

- Sell or not sell

- Positive or negative

- Accepted or rejected

Multi-class Classification

When the target or output variable contains more than two categories, it is referred to as multi-class classification. So, there can be multiple groups within the data, and every record can belong to any of the groups. For example:

- Yes or no or maybe

- Group A or group B or group C

- Category 1 or category 2 or category 3 or others

- Rank 1 or rank 2 or rank 3 or rank 4 or rank 5

Another useful property of supervised learning is that the model's performance can be evaluated on training and test data. Based on the type of model (classification or regression), the evaluation metric can be applied, and performance results can be measured. In this chapter, I will cover how to build machine learning models to execute regression and binary classification.

Building a Linear Regression Model

Linear regression refers to modeling the relationship between a set of independent variables and the output or dependent (numerical) variables. If the input variables include more than one variable, this is known as multivariable linear regression. In short, it is assumed that the dependent variable is a linear combination of other independent variables.

$$\bar{y} = B_0 + B_1 * X_1 + B_2 * X_{2+\ldots}$$

Here X_1, X_2, ... are the independent variables that are used to predict the output variable. The output of the linear regression is a straight line, which minimizes the actual vs. predicted values. A linear regression model cannot handle nonlinear data, as it's only possible to model linearly separable data, therefore polynomial regression is used for nonlinear data, and the output is generally a curve, instead of a straight line, as shown in Figure 6-4.

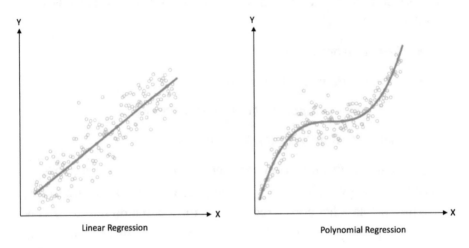

Figure 6-4. *Types of regression*

Linear regression also assumes that data is normally distributed, in order to improve prediction. Linear regression is one of the ways to predict continuous values, and you will see now how we can use other alternatives, to predict numerical output.

The following sections focus on solving regression tasks, using multiple machine learning algorithms. I will begin with data ingestion and exploratory data analysis and then build models. The steps 1 to 4 will be the same for all the regression models.

Note   Complete datasets, along with the relevant code, are available for reference from the GitHub repository for this book and execute best on Spark 2.3 and higher versions.

Let's build a linear regression model, using Spark's MLlib library, and predict the target variable, using the input features.

Reviewing the Data Information

The dataset that we are going to use for this example is a sample dataset and contains a total of 1,232 rows and 6 columns. We have to use 5 input variables to predict the target variable, using the linear regression model.

Step 1: Create the Spark Session Object

We start the Jupyter notebook, import SparkSession, and create a new SparkSession object to use with Spark.

```
[In]: from pyspark.sql import SparkSession
[In]: spark=SparkSession.builder.appName('supervised_ml').
getOrCreate()
```

Step 2: Read the Dataset

We then load and read the dataset within Spark, using DataFrame. We have to make sure that we have opened PySpark from the same directory folder where the dataset is available, or else we have to mention the directory path of the data folder.

```
[In]: df=spark.read.csv('Linear_regression_dataset.csv',infer
Schema=True,header=True)

[In]:print((df.count(), len(df.columns)))
[Out]: (1232, 6)
```

The preceding output confirms the size of our dataset, so we can then validate the datatypes of the input values, to check if we have to change/cast any column datatypes. In this example, all columns contain integer or double values that are already aligned with our requirements.

```
[In]: df.printSchema()
[Out]: root
 |-- var_1: integer (nullable = true)
 |-- var_2: integer (nullable = true)
 |-- var_3: integer (nullable = true)
 |-- var_4: double (nullable = true)
 |-- var_5: double (nullable = true)
 |-- output: double (nullable = true)
```

There is a total of six columns, of which five are input columns (var_1 to var_5) and target columns (label). We can now use the describe function to go over statistical measures of the dataset.

```
[In]: df.show(10)
[Out]:
+-----+-----+-----+-----+-----+-----+
|var_1|var_2|var_3|var_4|var_5|label|
+-----+-----+-----+-----+-----+-----+
|  734|  688|   81|0.328|0.259|0.418|
|  700|  600|   94| 0.32|0.247|0.389|
|  712|  705|   93|0.311|0.247|0.417|
|  734|  806|   69|0.315| 0.26|0.415|
|  613|  759|   61|0.302| 0.24|0.378|
|  748|  676|   85|0.318|0.255|0.422|
|  669|  588|   97|0.315|0.251|0.411|
|  667|  845|   68|0.324|0.251|0.381|
|  758|  890|   64| 0.33|0.274|0.436|
|  726|  670|   88|0.335|0.268|0.422|
+-----+-----+-----+-----+-----+-----+
only showing top 10 rows
```

Step 3: Feature Engineering

This is the part where we create a single vector combining all input features, by using Spark's VectorAssembler. It creates only a single feature that captures the input values for that particular row. So, instead of five input columns, the engine essentially translates the features into a single column with five input values, in the form of a list.

```
[In]: from pyspark.ml.linalg import Vector
[In]: from pyspark.ml.feature import VectorAssembler
```

We will pass all five input columns, to create a single features column.

```
[In]: df.columns
[Out]: ['var_1', 'var_2', 'var_3', 'var_4', 'var_5', 'label']

[In]: vec_assmebler=VectorAssembler(inputCols=['var_1',
'var_2', 'var_3', 'var_4', 'var_5'],outputCol='features')
[In]: features_df=vec_assmebler.transform(df)

[In]: features_df.printSchema()
[Out]: root
 |-- var_1: integer (nullable = true)
 |-- var_2: integer (nullable = true)
 |-- var_3: integer (nullable = true)
 |-- var_4: double (nullable = true)
 |-- var_5: double (nullable = true)
 |-- label: double (nullable = true)
 |-- features: vector (nullable = true)
```

As you can see, we have an additional column (features), which contains the single dense vector for all of the inputs. We then take a subset of the dataframe and select only the features column and the label column, to build the linear regression model.

```
[In]: df.select(['features','label']).show()
[Out]:
```

```
+--------------------+-----+
|            features|label|
+--------------------+-----+
|[734.0,688.0,81.0...|0.418|
|[700.0,600.0,94.0...|0.389|
|[712.0,705.0,93.0...|0.417|
|[734.0,806.0,69.0...|0.415|
|[613.0,759.0,61.0...|0.378|
|[748.0,676.0,85.0...|0.422|
|[669.0,588.0,97.0...|0.411|
|[667.0,845.0,68.0...|0.381|
|[758.0,890.0,64.0...|0.436|
|[726.0,670.0,88.0...|0.422|
|[583.0,794.0,55.0...|0.371|
|[676.0,746.0,72.0...|  0.4|
|[767.0,699.0,89.0...|0.433|
|[637.0,597.0,86.0...|0.374|
|[609.0,724.0,69.0...|0.382|
|[776.0,733.0,83.0...|0.437|
|[701.0,832.0,66.0...| 0.39|
|[650.0,709.0,74.0...|0.386|
|[804.0,668.0,95.0...|0.453|
|[713.0,614.0,94.0...|0.404|
+--------------------+-----+
only showing top 20 rows
```

Step 4: Split the Dataset

Let's split the dataset into training and test datasets, in order to train and evaluate the performance of the linear regression model. We split it according to a 70/30 ratio and train our model on 70% of the dataset. We can print the shape of the train and test data, to validate the size.

```
[In]: train, test = df.randomSplit([0.75, 0.25])
[In]:print(f"Size of train Dataset : {train.count()}" )
[Out]: Size of train Dataset : 911
[In]: print(f"Size of test Dataset : {test.count()}" )
[Out]:  Size of test Dataset : 321
```

Step 5: Build and Train Linear Regression Model

Now we build and train the linear regression model, using features, input, and label columns. We first import the linear regression from MLlib, as follows:

```
[In]: from pyspark.ml.regression import LinearRegression
```

```
[In]: lr = LinearRegression()
```

Note   For simplicity, all the machine learning models built in this chapter use default hyperparameters. Readers can use their own set of hyperparameters.

```
[In]:lr_model = lr.fit(train)
```

```
[In]: predictions_df=lr_model.transform(test)
```
```
[In]: predictions_df.show()
```

[Out]:

```
+-----+-----+-----+-----+-----+-----+--------------------+-------------------+
|var_1|var_2|var_3|var_4|var_5|label|            features|         prediction|
+-----+-----+-----+-----+-----+-----+--------------------+-------------------+
|  473|  499|   73|0.281|0.228|0.315|[473.0,499.0,73.0...| 0.3168228205906638|
|  498|  672|   61|0.288|0.238|0.325|[498.0,672.0,61.0...|0.33224574821552433|
|  513|  698|   61|0.298|0.236|0.339|[513.0,698.0,61.0...| 0.3314803948399759|
|  527|  569|   75|0.297|0.239|0.341|[527.0,569.0,75.0...| 0.3341226140464902|
|  532|  690|   69|0.303|0.245|0.351|[532.0,690.0,69.0...| 0.3399564421746622|
|  534|  609|   69|0.304|0.229|0.329|[534.0,609.0,69.0...|0.32848240366793496|
|  536|  531|   83|0.292|0.214|0.318|[536.0,531.0,83.0...|  0.328257290790113|
|  541|  830|   60|0.302|0.229| 0.33|[541.0,830.0,60.0...| 0.3418227186125283|
|  543|  615|   76|0.294|0.233|0.333|[543.0,615.0,76.0...|0.34119126494490326|
|  550|  631|   76|0.306|0.235|0.318|[550.0,631.0,76.0...| 0.3377949136360593|
|  550|  789|   54|0.305|0.238|0.359|[550.0,789.0,54.0...| 0.3439728742988891|
|  557|  659|   71|0.295|0.234|0.355|[557.0,659.0,71.0...|0.34713268891908144|
|  559|  613|   75|0.293|0.235|0.359|[559.0,613.0,75.0...| 0.34788019044256024|
|  569|  620|   77|0.302|0.247|0.349|[569.0,620.0,77.0...|0.35188383904315806|
|  569|  711|   65|0.305|0.237| 0.34|[569.0,711.0,65.0...| 0.3479171474465841|
|  570|  655|   66|0.311|0.246| 0.34|[570.0,655.0,66.0...| 0.3459593594270007|
|  570|  662|   73| 0.31|0.247|0.337|[570.0,662.0,73.0...| 0.3486539304568965|
|  570|  786|   57|0.301|0.249|0.366|[570.0,786.0,57.0...|0.35870641360119765|
|  572|  646|   71|0.311|0.235|0.329|[572.0,646.0,71.0...| 0.3419971659838297|
|  573|  634|   75|0.308|0.244|0.342|[573.0,634.0,75.0...|0.34846159793536235|
+-----+-----+-----+-----+-----+-----+--------------------+-------------------+
only showing top 20 rows
```

Step 6: Evaluate Linear Regression Model on Test Data

To check the performance of the model on unseen or test data, we make use of evaluate.

```
[In]: model_predictions=lr_model.evaluate(test)
[In]: model_predictions.r2
[Out]: 0.8855561089304634
[In]: print(model_predictions.meanSquaredError)
[Out]:0.00013305453514672318
```

Generalized Linear Model Regression

The generalized linear model (GLM) is an advanced version of linear regression that considers the target variable to have an error distribution other than a preferred normal distribution. The GLM generalizes linear regression, using a link function, so that variance is a function of the predicted value itself. Let's try to build the GLM on the same dataset and see if it performs better than a simple linear regression model. First, we must import the GLM from MLlib.

```
[In]: from pyspark.ml.regression import
GeneralizedLinearRegression
```

Step 1: Build and Train Generalized Linear Regression Model

```
[In]: glr = GeneralizedLinearRegression()
[In]: glr_model = glr.fit(train)

[In]: glr_model.coefficients

[Out]: DenseVector([0.0003, 0.0001, 0.0001, -0.6374, 0.4822])
```

We can get the coefficient values, using coefficient functions of that model. Here we can see that one of the features has a negative coefficient value. We can get more information about the GLM model, by using the summary function. It returns all the details, such as coefficient value, std error, AIC (Akaike information criterion) value, and p value.

[In]: glr_model.summary
[Out]:

```
Coefficients:
     Feature Estimate Std Error T Value P Value
 (Intercept)   0.1887    0.0169 11.1450  0.0000
       var_1   0.0003    0.0000 22.5404  0.0000
       var_2   0.0001    0.0000  4.3525  0.0000
       var_3   0.0002    0.0001  1.7469  0.0810
       var_4  -0.6258    0.0707 -8.8489  0.0000
       var_5   0.4610    0.0626  7.3697  0.0000

(Dispersion parameter for gaussian family taken to be 0.0001)
     Null deviance: 0.9886 on 905 degrees of freedom
 Residual deviance: 0.1355 on 905 degrees of freedom
AIC: -5429.3467
```

Step 2: Evaluate the Model Performance on Test Data

[In]: model_predictions=glr_model.evaluate(test)
[In]: model_predictions.predictions.show()

[Out]:

```
+-----+-----+-----+-----+-----+-----+--------------------+--------------------+
|var_1|var_2|var_3|var_4|var_5|label|            features|          prediction|
+-----+-----+-----+-----+-----+-----+--------------------+--------------------+
|  464|  640|   66|0.283| 0.22|0.301|[464.0,640.0,66.0...|  0.3141448302319312|
|  501|  774|   51|0.285|0.219|0.315|[501.0,774.0,51.0...|  0.3300426164818906|
|  533|  660|   62|0.296|0.233| 0.33|[533.0,660.0,62.0...| 0.33618025230208903|
|  534|  609|   69|0.304|0.229|0.329|[534.0,609.0,69.0...| 0.32776507532760846|
|  559|  613|   75|0.293|0.235|0.359|[559.0,613.0,75.0...| 0.34727124478074367|
|  562|  587|   80|0.308|0.235|0.344|[562.0,587.0,80.0...| 0.33805634511569305|
|  564|  648|   74|0.294|0.236|0.337|[564.0,648.0,74.0...|  0.3505501718299413|
|  568|  708|   57|0.311|0.247|0.347|[568.0,708.0,57.0...|  0.3471697543744474|
|  569|  544|   82|0.304| 0.24|0.343|[569.0,544.0,82.0...| 0.34339403169362914|
|  571|  577|   83|0.298|0.251|0.368|[571.0,577.0,83.0...| 0.35511010339509763|
|  573|  656|   75|0.313|0.242|0.345|[573.0,656.0,75.0...|  0.3449738829694534|
|  574|  556|   85|0.303|0.243|0.368|[574.0,556.0,85.0...| 0.34825256390850423|
|  575|  680|   68|  0.3|0.241|0.344|[575.0,680.0,68.0...|  0.3537472785088686|
|  576|  759|   57|0.313|0.254| 0.35|[576.0,759.0,57.0...|  0.3547252386784905|
|  578|  633|   76|0.309|0.249|0.337|[578.0,633.0,76.0...| 0.35151922591283996|
|  578|  733|   62|0.299|0.231|0.348|[578.0,733.0,62.0...| 0.35256482072466755|
|  579|  497|   91|0.304|0.225|0.352|[579.0,497.0,91.0...| 0.33834509029252136|
|  581|  724|   64|0.314|0.248|0.346|[581.0,724.0,64.0...| 0.35202791703249636|
|  582|  791|   52| 0.31| 0.24|0.359|[582.0,791.0,52.0...| 0.35293362472526907|
|  584|  680|   63|0.298|0.234| 0.35|[584.0,680.0,63.0...| 0.35400810480679434|
+-----+-----+-----+-----+-----+-----+--------------------+--------------------+
only showing top 20 rows
```

The Akaike information criterion (AIC) is an evaluation parameter of relative performance of quality of models for the same set dataset. AIC is mainly used to select among multiple models for a given dataset. A lesser value of AIC indicates that the model is of good quality. AIC tries to strike a balance between the variance and bias of the model. Therefore, it deals with the chances both of overfitting and underfitting. The model with the lowest AIC score is preferred over other models.

```
[In]: model_predictions.aic
[Out]: -1939.88
```

We can run the GLM for multiple distributions, such as

1. Binomial

2. Poisson

3. Gamma

4. Tweedie

```
[In]: glr = GeneralizedLinearRegression(family='Binomial')
[In]: glr_model = glr.fit(train)
[In]: model_predictions=glr_model.evaluate(test)
[In]: model_predictions.aic
[Out]: 336.991

[In]: glr = GeneralizedLinearRegression(family='Poisson')
[In]: glr_model = glr.fit(train)
[In]: predictions=glr_model.evaluate(test)
[In]: predictions.aic
[Out]: 266.53

[In]: glr = GeneralizedLinearRegression(family='Gamma')
[In]: glr_model = glr.fit(train)
[In]:model_predictions=glr_model.evaluate(test)
[In]: model_predictions.aic
[Out]: -1903.81
```

Here we can see that our default GLM model with Gaussian distribution has the lowest AIC value, compared to others.

Decision Tree Regression

The decision tree regression algorithm can be used for both regression and classification. It is quite powerful in terms of fitting the data well but comes with the high risk of sometimes overfitting the data. Decision trees contain multiple splits based on entropy or Gini indexes. The deeper the tree, the higher the chance of overfitting the data. In our example, we will build a decision tree for predicting the target value, with the default value of parameters (maxdepth = 5).

Step 1: Build and Train Decision Tree Regressor Model

```
[In]: from pyspark.ml.regression import DecisionTreeRegressor
[In]: dec_tree = DecisionTreeRegressor()
[In]: dec_tree_model = dec_tree.fit(train)
[In]: dec_tree_model.featureImportances
[Out]: SparseVector(5, {0: 0.9641, 1: 0.0193, 2: 0.0029, 3: 0.0053, 4: 0.0084})
```

Step 2: Evaluate the Model Performance on Test Data

```
[In]: model_predictions = dec_tree_model.transform(test)
[In]: model_predictions.show()
[Out]:
```

```
+-----+-----+-----+-----+-----+-----+--------------------+-------------------+
|var_1|var_2|var_3|var_4|var_5|label|            features|         prediction|
+-----+-----+-----+-----+-----+-----+--------------------+-------------------+
|  464|  640|   66|0.283| 0.22|0.301|[464.0,640.0,66.0...|            0.31925|
|  501|  774|   51|0.285|0.219|0.315|[501.0,774.0,51.0...|               0.33|
|  533|  660|   62|0.296|0.233| 0.33|[533.0,660.0,62.0...|               0.33|
|  534|  609|   69|0.304|0.229|0.329|[534.0,609.0,69.0...|            0.31925|
|  559|  613|   75|0.293|0.235|0.359|[559.0,613.0,75.0...|0.34612195121951217|
|  562|  587|   80|0.308|0.235|0.344|[562.0,587.0,80.0...|0.34612195121951217|
|  564|  648|   74|0.294|0.236|0.337|[564.0,648.0,74.0...|0.34612195121951217|
|  568|  708|   57|0.311|0.247|0.347|[568.0,708.0,57.0...|0.34612195121951217|
|  569|  544|   82| 0.24|0.343|[569.0,544.0,82.0...|0.34612195121951217|
|  571|  577|   83|0.298|0.251|0.368|[571.0,577.0,83.0...| 0.3534705882352937|
|  573|  656|   75|0.313|0.242|0.345|[573.0,656.0,75.0...|0.34612195121951217|
|  574|  556|   85|0.303|0.243|0.368|[574.0,556.0,85.0...|0.34612195121951217|
|  575|  680|   68|  0.3|0.241|0.344|[575.0,680.0,68.0...|0.34612195121951217|
|  576|  759|   57|0.313|0.254| 0.35|[576.0,759.0,57.0...|0.35724999999999996|
|  578|  633|   76|0.309|0.249|0.337|[578.0,633.0,76.0...| 0.3534705882352937|
|  578|  733|   62|0.299|0.231|0.348|[578.0,733.0,62.0...|0.35724999999999996|
|  579|  497|   91|0.304|0.225|0.352|[579.0,497.0,91.0...|0.34612195121951217|
|  581|  724|   64|0.314|0.248|0.346|[581.0,724.0,64.0...|0.35724999999999996|
|  582|  791|   52| 0.31| 0.24|0.359|[582.0,791.0,52.0...|0.35724999999999996|
|  584|  680|   63|0.298|0.234| 0.35|[584.0,680.0,63.0...|0.34612195121951217|
+-----+-----+-----+-----+-----+-----+--------------------+-------------------+
only showing top 20 rows
```

We import RegressionEvaluation from MLlib, to evaluate the performance of the decision tree on test data. As of now, there are two metrics available for evaluation: r^2 and RMSE (root mean squared error). r^2 mainly suggests how much of the variation in the dataset can be attributed to regression. Therefore, the higher the r^2, the better the performance of the model. On the other hand, RMSE suggests the total errors the model is making, in terms of the difference between actual and predicted values.

```
[In]: from pyspark.ml.evaluation import RegressionEvaluator
[In]: dt_evaluator = RegressionEvaluator(metricName='r2')
[In]: dt_r2 = dt_evaluator.evaluate(model_predictions)
[In]: print(f'The r-square value of DecisionTreeRegressor is {dt_r2}')
[Out]: The r-square value of DecisionTreeRegressor is 0.8093834699203476
[In]: dt_evaluator = RegressionEvaluator(metricName='rmse')
[In]: dt_rmse = dt_evaluator.evaluate(model_predictions)
[In]: print(f'The rmse value of DecisionTreeRegressor is {dt_rmse}')
[Out]: The rmse value of DecisionTreeRegressor is 0.014111932287681688
```

The r^2 value of this particular model is close to 0.81, which is a little lower than that of a simple linear regression model.

Random Forest Regressors

Random forest regressors are a collection of multiple individual decision trees built using different samples of data. The whole idea of combining these individual trees is to take majority voting or averages (in case of regression) to generalize effectively. A random forest is, therefore, an ensembling technique that takes a bagging approach. It can be used for regression as well as for classification tasks. Because decision trees tend to overfit the data, random forests remove the element of high variance, by taking the means of the predicted values from individual trees. In our example, we will build a random forest model for regression, using default parameters (numTrees = 20)

Step 1: Build and Train Random Forest Regressor Model

```
[In]: from pyspark.ml.regression import RandomForestRegressor

[In]: rf = RandomForestRegressor()

[In]: rf_model = rf.fit(train)
[In]:  rf_model.featureImportances
[Out]: SparseVector(5, {0: 0.4395, 1: 0.045, 2: 0.0243, 3: 0.2725, 4: 0.2188})
```

As you can see, the number of trees in the random forest is equal to 20. This number can be increased.

```
[In]: rf_model.getNumTrees
[Out]: 20
[In]: model_predictions = rf_model.transform(test)
[In]: model_predictions.show()
[Out]:
```

```
+-----+-----+-----+-----+-----+-----+--------------------+-------------------+
|var_1|var_2|var_3|var_4|var_5|label|            features|         prediction|
+-----+-----+-----+-----+-----+-----+--------------------+-------------------+
|  464|  640|   66|0.283| 0.22|0.301|[464.0,640.0,66.0...|0.32620528864346005|
|  501|  774|   51|0.285|0.219|0.315|[501.0,774.0,51.0...| 0.3315483384956547|
|  533|  660|   62|0.296|0.233| 0.33|[533.0,660.0,62.0...| 0.3279678481672696|
|  534|  609|   69|0.304|0.229|0.329|[534.0,609.0,69.0...|0.32849289839181284|
|  559|  613|   75|0.293|0.235|0.359|[559.0,613.0,75.0...|0.34431157381674565|
|  562|  587|   80|0.308|0.235|0.344|[562.0,587.0,80.0...|0.34633939901939004|
|  564|  648|   74|0.294|0.236|0.337|[564.0,648.0,74.0...|0.34660634825283587|
|  568|  708|   57|0.311|0.247|0.347|[568.0,708.0,57.0...| 0.3580627405761989|
|  569|  544|   82|0.304| 0.24|0.343|[569.0,544.0,82.0...| 0.3489929326622356|
|  571|  577|   83|0.298|0.251|0.368|[571.0,577.0,83.0...| 0.3538474256428915|
|  573|  656|   75|0.313|0.242|0.345|[573.0,656.0,75.0...|0.35044723951023526|
|  574|  556|   85|0.303|0.243|0.368|[574.0,556.0,85.0...|0.35245659207672475|
|  575|  680|   68|  0.3|0.241|0.344|[575.0,680.0,68.0...| 0.3508903073912221|
|  576|  759|   57|0.313|0.254| 0.35|[576.0,759.0,57.0...| 0.3667486309616494|
|  578|  633|   76|0.309|0.249|0.337|[578.0,633.0,76.0...|0.35079976893996123|
|  578|  733|   62|0.299|0.231|0.348|[578.0,733.0,62.0...| 0.3466773379187134|
|  579|  497|   91|0.304|0.225|0.352|[579.0,497.0,91.0...|0.34216301643346114|
|  581|  724|   64|0.314|0.248|0.346|[581.0,724.0,64.0...| 0.3584021380733537|
|  582|  791|   52| 0.31| 0.24|0.359|[582.0,791.0,52.0...| 0.3583885211293452|
|  584|  680|   63|0.298|0.234| 0.35|[584.0,680.0,63.0...| 0.350146054051407|
+-----+-----+-----+-----+-----+-----+--------------------+-------------------+
only showing top 20 rows
```

Step 2: Evaluate the Model Performance on Test Data

We can again use r^2 and RMSE as the evaluation parameter of the random forest model.

```
[In]:rf_evaluator = RegressionEvaluator(metricName='r2')
[In]: rf_r2 = rf_evaluator.evaluate(model_predictions)
[In]: print(f'The r-square value of RandomForestRegressor is
{rf_r2}')
[Out]: The r-square value of RandomForestRegressor is
0.8215863293044671
[In]: rf_evaluator = RegressionEvaluator(metricName='rmse')
[In]: rf_rmse = rf_evaluator.evaluate(model_predictions)
[In]: print(f'The rmse value of RandomForestRegressor is {rf_rmse}')
[Out]: The rmse value of RandomForestRegressor is
0.01365275410722947
```

As you can see, it clearly outperforms the decision tree regressor and has a higher r^2. The performance of this model can further be enhanced with hyperparameter tuning.

Gradient-Boosted Tree Regressor

A gradient-boosted tree (GBT) regressor is also an ensembling technique, which uses boosting under the hood. Boosting refers to making use of individual weak learners in order to boost the performance of the overall model. One major difference between bagging and boosting is that in bagging, the individual models that are built are parallel in nature, meaning they can be built independent of each other, but in boosting, the individual models are built in a sequential manner. In a gradient boosting approach, the second model focuses on the errors made by the first model and tries to reduce overall errors for those data points. Similarly, the next model tries to reduce the errors made by the previous model. In this way, the overall error of prediction is reduced. In the following example, we will build a GBT regressor with default parameters.

Step 1: Build and Train a GBT Regressor Model

```
[In]: from pyspark.ml.regression import GBTRegressor
[In]: gbt = GBTRegressor()
[In]: gbt_model=gbt.fit(train)
[In]: gbt_model.featureImportances
[Out]: SparseVector(5, {0: 0.2325, 1: 0.2011, 2: 0.1645, 3:
0.2268, 4: 0.1751})
[In]: model_predictions = gbt_model.transform(test)
[In]: model_predictions.show()
[Out]:
```

```
+-----+-----+-----+-----+-----+-----+--------------------+-------------------+
|var_1|var_2|var_3|var_4|var_5|label|            features|         prediction|
+-----+-----+-----+-----+-----+-----+--------------------+-------------------+
|  464|  640|   66|0.283| 0.22|0.301|[464.0,640.0,66.0...|0.31256952473048055|
|  501|  774|   51|0.285|0.219|0.315|[501.0,774.0,51.0...| 0.3264479872682094|
|  533|  660|   62|0.296|0.233| 0.33|[533.0,660.0,62.0...|0.33244918944928964|
|  534|  609|   69|0.304|0.229|0.329|[534.0,609.0,69.0...| 0.3152040911562068|
|  559|  613|   75|0.293|0.235|0.359|[559.0,613.0,75.0...| 0.3460779548096019|
|  562|  587|   80|0.308|0.235|0.344|[562.0,587.0,80.0...| 0.3462514921128048|
|  564|  648|   74|0.294|0.236|0.337|[564.0,648.0,74.0...|0.33944147594999274|
|  568|  708|   57|0.311|0.247|0.347|[568.0,708.0,57.0...| 0.3430537197909264|
|  569|  544|   82|0.304| 0.24|0.343|[569.0,544.0,82.0...| 0.3438444048425846|
|  571|  577|   83|0.298|0.251|0.368|[571.0,577.0,83.0...| 0.3644713166785006|
|  573|  656|   75|0.313|0.242|0.345|[573.0,656.0,75.0...|0.33886550037240126|
|  574|  556|   85|0.303|0.243|0.368|[574.0,556.0,85.0...|0.35592388197078745|
|  575|  680|   68|  0.3|0.241|0.344|[575.0,680.0,68.0...|0.34563903926814665|
|  576|  759|   57|0.313|0.254| 0.35|[576.0,759.0,57.0...| 0.3539939919952057|
|  578|  633|   76|0.309|0.249|0.337|[578.0,633.0,76.0...|0.35539649979287086|
|  578|  733|   62|0.299|0.231|0.348|[578.0,733.0,62.0...|0.35762532040568573|
|  579|  497|   91|0.304|0.225|0.352|[579.0,497.0,91.0...| 0.3399451444570211|
|  581|  724|   64|0.314|0.248|0.346|[581.0,724.0,64.0...| 0.3551771577061834|
|  582|  791|   52| 0.31| 0.24|0.359|[582.0,791.0,52.0...| 0.3537833213771059|
|  584|  680|   63|0.298|0.234| 0.35|[584.0,680.0,63.0...| 0.3457064089681957|
+-----+-----+-----+-----+-----+-----+--------------------+-------------------+
only showing top 20 rows
```

Step 2: Evaluate the Model Performance on Test Data

```
[In]: gbt_evaluator = RegressionEvaluator(metricName='r2')
[In]: gbt_r2 = gbt_evaluator.evaluate(model_predictions)
[In]: print(f'The r-square value of GradientBoostedRegressor is
{gbt_r2}')
[Out]: The r-square value of GradientBoostedRegressor is
0.8477273892307596
[In]: gbt_evaluator = RegressionEvaluator(metricName='rmse')
[In]: gbt_rmse = gbt_evaluator.evaluate(model_predictions)
[In]: print(f'The rmse value of GradientBoostedRegressor is
{gbt_rmse}')
[Out]: The rmse value of GradientBoostedRegressor is
0.013305445803592103
```

As you can see, the GBT regressor outperforms the random forestmodel. With r^2 being close to 0.85, it can be considered the final model, after proper tuning.

Building Multiple Models for Binary Classification Tasks

In this third and final section of the chapter, you will see how to build multiple machine learning models for binary classification tasks. The data that we are going to use for this is a subset of an open source Bank Marketing Data Set from the UCI ML repository, available at `https://archive.ics.uci.edu/ml/datasets/Bank+Marketing`.

There are two reasons for selecting only a subset of this data. The first is to maintain the class balance for the classification task, so as not to make it an anomalous detection category task. Another reason for selecting only a subset of the features is to limit the amount of signals in the data, as some of the features in the dataset strongly affect the output and, therefore, are ignored in this exercise.

The dataset contains 9,500 rows and 8 columns. The idea is to predict if the user will subscribe to another product or service (term deposit), based on the other attributes, such as age, job, loan, etc. This is a typical requirement in which machine learning is leveraged to find the top users who can be targeted by the business for cross-selling or upselling.

I'll begin with the logistic regression model.

Logistic Regression

Logistic regression is considered to be one of baseline models, owing to its simplicity and interpretability. Under the hood, it is quite similar to linear regression. It also assumes that output is a linear combination of the dependent variables, but to keep the output between 0 and 1, as it returns the probability as output, it makes use of a nonlinear function (sigmoid), which produces an S curve instead of a straight line (linear regression).

We'll start by building the baseline in Steps 1–3 and then complete the logistic regression model with default hyperparameters, in Steps 4–6.

Step 1: Read the Dataset

```
[In]: df=spark.read.csv('bank_data.csv',inferSchema=True,
header=True)
[In]: df.count()
[Out]: 9501

[In]: df.columns
[Out]: ['age', 'job', 'marital', 'education', 'default',
'housing','loan', 'target_class']
[In]: df.printSchema()
[Out]:
```

```
root
 |-- age: integer (nullable = true)
 |-- job: string (nullable = true)
 |-- marital: string (nullable = true)
 |-- education: string (nullable = true)
 |-- default: string (nullable = true)
 |-- housing: string (nullable = true)
 |-- loan: string (nullable = true)
 |-- target_class: string (nullable = true)
```

As you can see, the input columns are all the columns, except for the target class column. The target class is also well-balanced, in terms of the count of yes and no labels. We will have to convert yeses and noes into 1s and 0s, as well as rename the target_class column to "label," which is the default acceptance column name in machine learning model parameters.

```
[In]: df.groupBy('target_class').count().show()
[Out]:
```

```
+-------------+-----+
|target_class|count|
+-------------+-----+
|           no| 4861|
|          yes| 4640|
+-------------+-----+
```

Step 2: Feature Engineering for Model

```
[In]: from pyspark.sql import functions as F
[In]: from pyspark.sql import *
[In]: df=df.withColumn("label", F.when(df.target_class =='no',
F.lit(0)).otherwise(F.lit(1)))
[In]: df.groupBy('label').count().show()
[Out]:
```

```
+-----+-----+
|label|count|
+-----+-----+
|    1| 4640|
|    0| 4861|
+-----+-----+
```

Now that we have renamed the output column "label" and converted the target class to 1s and 0s, the next step is to create features for the model. Because we have categorical columns, such as job and edu, we will have to use StringIndexer and OneHotEncoder to convert them into a numerical format. We create a Python function, cat_to_num, to convert all the categorical features into numerical ones.

```
[In]: from pyspark.ml.feature import OneHotEncoder,
StringIndexer, VectorAssembler
[In]: def cat_to_num(df):

    for col in df.columns:
        stringIndexer = StringIndexer(inputCol=col,
        outputCol=col+"_index")
```

```
    model = stringIndexer.fit(df)
    indexed = model.transform(df)
    encoder = OneHotEncoder(inputCol=col+"_index",
    outputCol=col+"_vec")
    df = encoder.transform(indexed)
  df_assembler = VectorAssembler(inputCols=['age','marit
  al_vec','education_vec','default_vec','housing_vec','loan_
  vec'], outputCol="features")
  df = df_assembler.transform(df)
  return df.select(['features','label'])
```

We just select the new features column and target label column, as we don't need the earlier original columns for model training.

```
[In]: df_new=cat_to_num(df)
[In]: df_new.show()
[Out]:
```

```
+--------------------+-----+
|            features|label|
+--------------------+-----+
|(16,[0,1,8,11,13,...|    0|
|(16,[0,1,5,13,14]...|    0|
|(16,[0,1,5,11,12,...|    0|
|(16,[0,1,9,11,13,...|    0|
|(16,[0,1,5,11,13,...|    0|
|(16,[0,1,6,13,14]...|    0|
|(16,[0,1,7,11,13,...|    0|
|(16,[0,1,10,13,14...|    0|
|(16,[0,2,7,11,12,...|    0|
|(16,[0,2,5,11,12,...|    0|
|(16,[0,1,10,13,14...|    0|
|(16,[0,2,5,11,12,...|    0|
|(16,[0,2,5,11,13,...|    0|
|(16,[0,3,8,11,12,...|    0|
|(16,[0,1,9,11,12,...|    0|
|(16,[0,1,6,12,15]...|    0|
|(16,[0,1,9,11,12,...|    0|
|(16,[0,1,9,12,15]...|    0|
|(16,[0,1,6,11,12,...|    0|
|(16,[0,2,6,13,14]...|    0|
+--------------------+-----+
only showing top 20 rows
```

Now we have all the input features merged into a single dense vector ('features'), along with output column labels, which we can use to train the machine learning models. The new dataframe created using only two columns (features, label) is now called df_new and will be used for every model. We can now split this new dataframe into train and test datasets. We can split the data into a 75%/25% ratio, using the randomplit function.

Step 3: Split the Data into Train and Test Datasets

```
[In]: train, test = df_new.randomSplit([0.75, 0.25])
[In]: print(f"Size of train Dataset : {train.count()}" )
[Out]: 7121
[In]: print(f"Size of test Dataset : {test.count()}" )
[Out]: 2380
```

Step 4: Build and Train the Logistic Regression Model

```
[In]: from pyspark.ml.classification import LogisticRegression
[In]: lr = LogisticRegression()
[In]: lr_model = lr.fit(train)
[In]:print( lr_model.coefficients)
[Out]:
[0.0272019114172,-0.647672064875,0.229030508111,-
0.77074788287,-12.36869511,-12.8865599132,-
13.2257790609,-12.6705131313,-13.0023164274,-13.074766258-
6,-12.6985757761,1.42220523957,0.301582233094,-
0.0127231892838,0.218471149577,0.332362933568]
```

Once the model is built, we can make use of the internal function summary, which offers important details regarding the model, such as ROC curve, precision, recall, AUC (area under the curve), etc.

Step 5: Evaluate Performance on Training Data

```
[In]: lr_summary=lr_model.summary
[In]: lr_summary.accuracy
[Out]: 0.673079623648364
[In]: lr_summary.areaUnderROC
[Out]: 0.7186044694483512
[In]: lr_summary.weightedRecall
[Out]: 0.673079623648364
[In]: lr_summary.weightedPrecision
[Out]: 0.6750967624018298
```

Here, using the summary function, we can view the model's performance on train data, such as its accuracy, AUC, weighted recall, and precision. We can also view additional details—such as how precision varies for various threshold values, the relation between precision and recall, and how recall varies with different threshold values—to pick the right threshold value for the model. These also can be plotted, to view the relationships.

```
[In]: lr_summary.precisionByThreshold.show()
[Out]:
```

```
+------------------+------------------+
|         threshold|         precision|
+------------------+------------------+
|0.9999970343745905|0.8888888888888888|
|0.8228312020653131|0.7582417582417582|
|0.7967647456585993|0.7207792207792207|
| 0.781075991097264|0.7753623188405797|
|0.7651034811582511|0.7657142857142857|
|0.7542442431736941| 0.748868778280543|
|0.7449881650355308|0.7350096711798839|
|0.7347410924161487|0.7302631578947368|
|0.7247619936276426|0.7335285505124451|
|0.7154742177815616|0.7370466321243523|
|0.7087858642292815|0.7197231833910035|
|0.7025970714160398|0.7222808870116156|
|0.6952841089393125|0.7245681381957774|
|0.6880881359369632|0.7221727515583259|
|0.6811170847960969| 0.725328947368421|
|0.6744212577411599|0.7251552795031055|
|0.6686109758051513|0.7203513909224012|
|0.6619238394946346|0.7225716282320056|
|0.6564481313737397|0.7228915662650602|
|0.6508779720961013|0.7238278741168914|
+------------------+------------------+
only showing top 20 rows
```

[In]: lr_summary.roc.show()
[Out]:

```
+--------------------+--------------------+
|                 FPR|                 TPR|
+--------------------+--------------------+
|                 0.0|                 0.0|
|0.001097393689986...|0.009205983889528193|
|0.006035665294924554|0.019850402761795168|
|0.011796982167352537| 0.03193325661680092|
| 0.0170096021947838|0.061565017261219795|
|0.022496570644718793| 0.07710011507479862|
|0.030452674897119343| 0.09522439585730726|
| 0.03758573388203018|  0.1093210586881473|
|0.044993141289437585|  0.1277330264672037|
| 0.04993141289437586|  0.1441311852704258|
| 0.05569272976680384|  0.1636939010356732|
| 0.06666666666666667| 0.17951668584579977|
| 0.07215363511659809| 0.19677790563866512|
| 0.07873799725651577|  0.2172036823935558|
| 0.08559670781893004| 0.23331415420023016|
|  0.0916323731138546|  0.2537399309551208|
|  0.097119341563786| 0.26869965477560415|
|    0.10480109739369| 0.28308400460299193|
| 0.10891632373113855|  0.2974683544303797|
| 0.11358024691358025| 0.31070195627157654|
+--------------------+--------------------+
only showing top 20 rows
```

[In]: lr_summary.recallByThreshold.show()
[Out]:

```
+------------------+--------------------+
|         threshold|              recall|
+------------------+--------------------+
|0.9999970343745905|0.009205983889528193|
|0.8228312020653131|0.019850402761795168|
|0.7967647456585993| 0.03193325661680092|
| 0.781075991097264|0.061565017261219795|
|0.7651034811582511| 0.07710011507479862|
|0.7542442431736941| 0.09522439585730726|
|0.7449881650355308|  0.1093210586881473|
|0.7347410924161487|  0.1277330264672037|
|0.7247619936276426|  0.1441311852704258|
|0.7154742177815616|  0.1636939010356732|
|0.7087858642292815| 0.17951668584579977|
|0.7025970714160398| 0.19677790563866512|
|0.6952841089393125|  0.2172036823935558|
|0.6880881359369632| 0.23331415420023016|
|0.6811170847960969|  0.2537399309551208|
|0.6744212577411599| 0.26869965477560415|
|0.6686109758051513| 0.28308400460299193|
|0.6619238394946346|  0.2974683544303797|
|0.6564481313737397| 0.31070195627157654|
|0.6508779720961013| 0.32422324510932105|
+------------------+--------------------+
only showing top 20 rows
```

```
[In]: lr_summary.pr.show()
[Out]:
```

```
+--------------------+------------------+
|              recall|         precision|
+--------------------+------------------+
|                 0.0|0.8888888888888888|
|0.009205983889528193|0.8888888888888888|
|0.019850402761795168|0.7582417582417582|
| 0.03193325661680092|0.7207792207792207|
|0.061565017261219795|0.7753623188405797|
| 0.07710011507479862|0.7657142857142857|
| 0.09522439585730726| 0.748868778280543|
|  0.1093210586881473|0.7350096711798839|
|  0.1277330264672037|0.7302631578947368|
|  0.1441311852704258|0.7335285505124451|
|  0.1636939010356732|0.7370466321243523|
| 0.17951668584579977|0.7197231833910035|
| 0.19677790563866512|0.7222808870116156|
|  0.2172036823935558|0.7245681381957774|
| 0.23331415420023016|0.7221727515583259|
|  0.2537399309551208| 0.725328947368421|
| 0.26869965477560415|0.7251552795031055|
| 0.28308400460299193|0.7203513909224012|
|  0.2974683544303797|0.7225716282320056|
| 0.31070195627157654|0.7228915662650602|
+--------------------+------------------+
only showing top 20 rows
```

Step 6: Evaluate Performance on Test Data

```
[In]: model_predictions = lr_model.transform(test)
[In]: model_predictions.columns
[Out]: ['features', 'label', 'rawPrediction', 'probability',
'prediction']
[In]: model_predictions.select(['label','probability',
'prediction']).show(10,False)
[Out]:
```

```
+-----+---------------------------------------------+----------+
|label|probability                                  |prediction|
+-----+---------------------------------------------+----------+
|1    |[0.7921142321252156,0.2078857678747844]      |0.0       |
|0    |[0.7332732992042076,0.2667267007957924]      |0.0       |
|0    |[0.7058440774094991,0.294155922590501]       |0.0       |
|0    |[0.5801868096061101,0.41981319039388987]     |0.0       |
|1    |[0.5735471647731092,0.42645283522689076]     |0.0       |
|1    |[0.5601903397477959,0.4398096602522042]      |0.0       |
|0    |[0.5332326656582489,0.46676733434175105]     |0.0       |
|1    |[0.5332326656582489,0.46676733434175105]     |0.0       |
|1    |[0.5196706190017784,0.48032938099822164]     |0.0       |
|1    |[0.4585716734035034,0.5414283265964965]      |1.0       |
+-----+---------------------------------------------+----------+
only showing top 10 rows
```

As you can see, the prediction column shows the model prediction for each of the records in the test data. The probability column shows the values for both classes (0 & 1). The probability at 0th index is of 0; the other is for a prediction of 1. The evaluation of the logistic regression model on test data can be done using BinaryClassEvaluator. We can get the area under ROC and that under the PR curve, as shown following:

```
[In]:from pyspark.ml.evaluation import
BinaryClassificationEvaluator
[In]:lr_evaluator = BinaryClassificationEvaluator(metricName=
'areaUnderROC')
[In]: lr_auroc = lr_evaluator.evaluate(model_predictions)
[In]: print(f'The auroc value of Logistic Regression Model is
{lr_auroc}')
[Out]: The auroc value of Logistic Regression Model is
0.7092938229110143

[In]: lr_evaluator = BinaryClassificationEvaluator(metricName=
'areaUnderPR')
[In]: lr_aupr = lr_evaluator.evaluate(model_predictions)
[In]: print(f'The aupr value of Logistic Regression Model is
{lr_aupr}')
```

```
[Out]: The aupr value of Logistic Regression Model is
0.6630743130940658

[In]: true_pos=model_predictions.filter(model_
predictions['label']==1).filter(model_
predictions['prediction']==1).count()

[In]: actual_pos=model_predictions.filter(model_
predictions['label']==1).count()
[In]: pred_pos=model_predictions.filter(model_
predictions['prediction']==1).count()
Recall
[In]: float(true_pos)/(actual_pos)
[Out]: 0.6701030927835051
Precision
[In]: float(true_pos)/(pred_pos)
[Out]: 0.6478405315614618
```

Decision Tree Classifier

As mentioned earlier, decision trees can be used for classification
as well as regression. Here, we will build a decision tree with default
hyperparameters and use it to predict whether the user will opt for the new
term deposit plan.

Step 1: Build and Train Decision Tree Classifier Model

```
[In]: from pyspark.ml.classification import
DecisionTreeClassifier
[In]: dt = DecisionTreeClassifier()
[In]: dt_model = dt.fit(train)
[In]: model_predictions = dt_model.transform(test)
```

```
[Out]: model_predictions.select(['label','probability',
'prediction']).show(10,False)
```

```
+-----+--------------------------------------------+----------+
|label|probability                                 |prediction|
+-----+--------------------------------------------+----------+
|1    |[0.8089887640449438,0.19101123595505617]    |0.0       |
|0    |[0.8089887640449438,0.19101123595505617]    |0.0       |
|0    |[0.8089887640449438,0.19101123595505617]    |0.0       |
|0    |[0.3055555555555556,0.6944444444444444]     |1.0       |
|1    |[0.3055555555555556,0.6944444444444444]     |1.0       |
|1    |[0.3055555555555556,0.6944444444444444]     |1.0       |
|0    |[0.42972350230414746,0.5702764976958525]    |1.0       |
|1    |[0.42972350230414746,0.5702764976958525]    |1.0       |
|1    |[0.42972350230414746,0.5702764976958525]    |1.0       |
|1    |[0.42972350230414746,0.5702764976958525]    |1.0       |
+-----+--------------------------------------------+----------+
only showing top 10 rows
```

Step 2: Evaluate Performance on Test Data

```
[In]: dt_evaluator = BinaryClassificationEvaluator(metricName=
'areaUnderROC')
[In]: dt_auroc = dt_evaluator.evaluate(model_predictions)
[In]: print(f'The auc value of Decision Tree Classifier Model
is {dt_auroc}')
[Out]: The auc value of Decision Tree Classifier Model is
0.516199386190993
[In]: dt_evaluator = BinaryClassificationEvaluator(metricName=
'areaUnderPR')
[In]: dt_aupr = dt_evaluator.evaluate(model_predictions)
[In]: print(f'The aupr value of Decision Tree Model is {dt_aupr}')
[Out]: The aupr value of Decision Tree Model is
0.46771834172588167

[In]: true_pos=model_predictions.filter(model_
predictions['label']==1).filter(model_
predictions['prediction']==1).count()
```

```
[In]: actual_pos=model_predictions.filter(model_
predictions['label']==1).count()
```

```
[In]: pred_pos=model_predictions.filter(model_
predictions['prediction']==1).count()
```

```
[In]: float(true_pos)/(actual_pos)
[Out]: 0.6907216494845361
[In]: float(true_pos)/(pred_pos)
[Out]: 0.6661143330571665
```

Support Vector Machines Classifiers

Support vector machines (SVMs) are used for classification tasks, as they find the hyperplane that maximizes the margin (perpendicular distance) between two classes. All the instances and target classes are represented as vectors in high-dimensional space, and the SVM finds the closest two points from the two classes that support the best separating line or hyperplane, as shown in Figure 6-5.

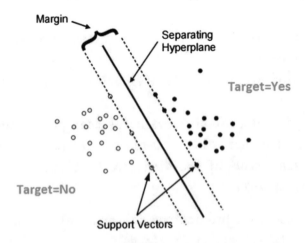

Figure 6-5. *Support vector machine*

For nonlinearly separable data, there are different kernel tricks to separate the classes. In our example, we will build a linearly separable support vector classifier with default hyperparameters.

Step 1: Build and Train SVM Model

```
[In]: from pyspark.ml.classification import LinearSVC
[In]: lsvc = LinearSVC()
[In]: lsvc_model = lsvc.fit(train)
[In]: model_predictions = lsvc_model.transform(test)
[In]: model_predictions.columns
[Out]: ['features', 'label', 'rawPrediction', 'prediction']
[In]:model_predictions.select(['label','prediction']).show(10,False)
[Out]:
```

```
+-----+----------+
|label|prediction|
+-----+----------+
|1    |0.0       |
|0    |0.0       |
|0    |0.0       |
|0    |1.0       |
|1    |1.0       |
|1    |1.0       |
|0    |1.0       |
|1    |1.0       |
|1    |1.0       |
|1    |1.0       |
+-----+----------+
only showing top 10 rows
```

Step 2: Evaluate Performance on Test Data

```
[In]: svc_evaluator = BinaryClassificationEvaluator(metricName=
'areaUnderROC')
[In]: svc_auroc = svc_evaluator.evaluate(model_predictions)
```

```
[In]: print(f'The auc value of SupportVectorClassifier  is
{svc_auroc}')
[Out]: The auc value of SupportVectorClassifier  is
0.7043772749366973

[In]: svc_evaluator = BinaryClassificationEvaluator(metricName=
'areaUnderPR')
[In]: svc_aupr =svc_evaluator.evaluate(model_predictions)
[In]: print(f'The aupr value of SupportVectorClassifier Model
is {svc_aupr}')
[Out]: The aupr value of SupportVectorClassifier Model is
0.6567277377856992

[In]: true_pos=model_predictions.filter(model_
predictions['label']==1).filter(model_
predictions['prediction']==1).count()
[In]: actual_pos=model_predictions.filter(model_
predictions['label']==1).count()
[In]: pred_pos=model_predictions.filter(model_
predictions['prediction']==1).count()
[In]: float(true_pos)/(actual_pos)
[Out]: 0.7774914089347079

[In]: float(true_pos)/(pred_pos)
[Out]: 0.600132625994695
```

Naive Bayes Classifier

Naive Bayes (NB) classifiers work on the principle of conditional probability and assume absolute independence between predictors. An NB classifier doesn't have many hyperparameters and can outperform some of the most sophisticated algorithms out there. In the following example, we will build an NB classifier and evaluate its performance on the test data.

Step 1: Build and Train SVM Model

```
[In]: from pyspark.ml.classification import NaiveBayes
[In]: nb = NaiveBayes()
[In]: nb_model = nb.fit(train)
[In]: model_predictions = nb_model.transform(test)
[In]: model_predictions.select(['label','probability',
'prediction']).show(10,False)
[Out]:
```

```
+-----+----------------------------------------+----------+
|label|probability                             |prediction|
+-----+----------------------------------------+----------+
|1    |[0.48630071280188164,0.5136992871981184]|1.0       |
|0    |[0.49911662267320056,0.5008833773267994]|1.0       |
|0    |[0.5044578946407129,0.49554210535928717]|0.0       |
|0    |[0.40355230543479254,0.5964476945652075]|1.0       |
|1    |[0.4045812569659479,0.5954187430340521] |1.0       |
|1    |[0.4066416699033005,0.5933583300966995] |1.0       |
|0    |[0.41077229948270594,0.589227700517294] |1.0       |
|1    |[0.41077229948270594,0.589227700517294] |1.0       |
|1    |[0.41284238036891824,0.5871576196310817]|1.0       |
|1    |[0.42219493600486624,0.5778050639951338]|1.0       |
+-----+----------------------------------------+----------+
only showing top 10 rows
```

Step 2: Evaluate Performance on Test Data

```
[In]: nb_evaluator = BinaryClassificationEvaluator(metricName='
areaUnderROC')
[In]: nb_auroc = nb_evaluator.evaluate(model_predictions)
[In]: print(f'The auc value of NB Classifier is {nb_auroc}')
[Out]: The auc value of NB Classifier is 0.43543736717760884

[In]: nb_evaluator = BinaryClassificationEvaluator(metricName='
areaUnderPR')
[In]: nb_aupr =nb_evaluator.evaluate(model_predictions)
[In]: print(f'The aupr value of NB Classifier Model is {nb_aupr}')
[Out]: The aupr value of NB Classifier Model is 0.4321001351769349
```

```
[In]: true_pos=model_predictions.filter(model_
predictions['label']==1).filter(model_
predictions['prediction']==1).count()
[In]: actual_pos=model_predictions.filter(model_
predictions['label']==1).count()
[In]: pred_pos=model_predictions.filter(model_
predictions['prediction']==1).count()

[In]: float(true_pos)/(actual_pos)
[Out]: 0.586
[In]: float(true_pos)/(pred_pos)
[Out]: 0.625
```

Gradient Boosted Tree Classifier

So far, we have used single algorithms for classification. Now we move on to use ensemble methods, such as GBT and random forests, for classification. Bagging and boosting for classification works according to similar principles as regression.

Step 1: Build and Train the GBT Model

```
[In]: from pyspark.ml.classification import GBTClassifier
[In]: gbt = GBTClassifier()
[In]: gbt_model = gbt.fit(train)
[In]: model_predictions = gbt_model.transform(test)
[In]: model_predictions.select(['label','probability',
'prediction']).show(10,False)
[Out]:
```

```
+-----+------------------------------------------------+----------+
|label|probability                                     |prediction|
+-----+------------------------------------------------+----------+
|1    |[0.8118906324166552,0.18810936758334484]|0.0       |
|0    |[0.8045448087863976,0.19545519121360244]|0.0       |
|0    |[0.7451850305979334,0.2548149694020666] |0.0       |
|0    |[0.30631795902192194,0.6936820409780781]|1.0       |
|1    |[0.31352609339453574,0.6864739066054643]|1.0       |
|1    |[0.3281403432684641,0.6718596567315359] |1.0       |
|0    |[0.5130658281231842,0.48693417187681576]|0.0       |
|1    |[0.5130658281231842,0.48693417187681576]|0.0       |
|1    |[0.48461420867158717,0.5153857913284128]|1.0       |
|1    |[0.4717208467161091,0.5282791532838909] |1.0       |
+-----+------------------------------------------------+----------+
only showing top 10 rows
```

Step 2: Evaluate Performance on Test Data

```
[In]: gbt_evaluator = BinaryClassificationEvaluator(metricName=
'areaUnderROC')
[In]: gbt_auroc = gbt_evaluator.evaluate(model_predictions)
[In]: print(f'The auc value of GradientBoostedTreesClassifier
is {gbt_auroc}')
[Out]: The auc value of GradientBoostedTreesClassifier is
0.7392410330756018

[In]: gbt_evaluator = BinaryClassificationEvaluator(metricName=
'areaUnderPR')
[In]: gbt_aupr = gbt_evaluator.evaluate(model_predictions)
[In]: print(f'The aupr value of GradientBoostedTreesClassifier
Model is {gbt_aupr}')
[Out]: The aupr value of GradientBoostedTreesClassifier Model
is 0.7345982892755392

[In]: true_pos=model_predictions.filter(model_
predictions['label']==1).filter(model_
predictions['prediction']==1).count()
```

```
[In]: actual_pos=model_predictions.filter(model_
predictions['label']==1).count()
[In]: pred_pos=model_predictions.filter(model_
predictions['prediction']==1).count()
[In]: float(true_pos)/(actual_pos)
[Out]: 0.668
[In]: float(true_pos)/(pred_pos)
[Out]: 0.674
```

Random Forest Classifier

Once again, a random forest classifier is a collection of multiple decision tree classifiers. It works on the voting mechanism and predicts the output class that received the maximum votes from all individual decision trees. Let's build a random forest classifier with the same data.

Step 1: Build and Train the Random Forest Model

```
[In]: from pyspark.ml.classification import
RandomForestClassifier
[In]: rf = RandomForestClassifier(numTrees=50,maxDepth=30)
[In]: rf_model = rf.fit(train)
[In]: model_predictions=rf_model.transform(test)
[In]: model_predictions.select(['label','probability',
'prediction']).show(10,False)
[Out]:
```

```
+-----+---------------------------------------------+----------+
|label|probability                                  |prediction|
+-----+---------------------------------------------+----------+
|1    |[0.6587687671584254,0.3412312328415746]      |0.0       |
|0    |[0.6662919718482153,0.3337080281517847]      |0.0       |
|0    |[0.6356045667482568,0.36439543325174306]     |0.0       |
|0    |[0.4585219274408326,0.5414780725591675]      |1.0       |
|1    |[0.4585219274408326,0.5414780725591675]      |1.0       |
|1    |[0.4596191735681992,0.5403808264318009]      |1.0       |
|0    |[0.4627248123254553,0.5372751876745447]      |1.0       |
|1    |[0.4627248123254553,0.5372751876745447]      |1.0       |
|1    |[0.4627248123254553,0.5372751876745447]      |1.0       |
|1    |[0.467969181062204,0.532030818937796]        |1.0       |
+-----+---------------------------------------------+----------+
only showing top 10 rows
```

Step 2: Evaluate Performance on Test Data

[In]: rf_evaluator = BinaryClassificationEvaluator(metricName=' areaUnderROC')

[In]: rf_auroc = rf_evaluator.evaluate(model_predictions)

[In]: print(f'The auc value of RandomForestClassifier Model is {rf_auroc}')

[Out]:

The auc value of RandomForestClassifier Model is

0.7326433634020617

[In]: rf_evaluator = BinaryClassificationEvaluator(metricName=' areaUnderPR')

[In]: rf_aupr = rf_evaluator.evaluate(model_predictions)

[In]: print(f'The aupr value of RandomForestClassifier Model is {rf_aupr}')

[Out]: The aupr value of RandomForestClassifier Model is

0.7277253895494864

[In]; true_pos=model_predictions.filter(model_ predictions['label']==1).filter(model_ predictions['prediction']==1).count()

```
[In]: actual_pos=model_predictions.filter(model_
predictions['label']==1).count()
[In]: pred_pos=model_predictions.filter(model_
predictions['prediction']==1).count()
[In]: float(true_pos)/(actual_pos)
[Out]: 0.67
[In]: float(true_pos)/(pred_pos)
[Out]: 0.67
```

So far, we have been using the default hyperparameters for all the models, but it's rarely the case that the model can perform to the best of its ability with those default settings. Therefore, it's imperative that we tune the models for the right combination of hyperparameters.

Hyperparameter Tuning and Cross-Validation

In the following example, we will take the random forest model that we just built and try to find the best combination of its hyperparameters, to improve performance. We can use `ParamgridBuilder` and `CrossValidator` for hyperparameter tuning. We will pass different values in the parameter grid for three hyperparameters (`maxDepth`, `maxBins`, and `numTrees`). It might take a few minutes to complete, as it builds random forest models for all of these combinations before returning the best possible combination for this train dataset.

```
[In]: from pyspark.ml.tuning import ParamGridBuilder,
CrossValidator
[In]: rf = RandomForestClassifier()
[In]: paramGrid = (ParamGridBuilder()
              .addGrid(rf.maxDepth, [5,10,20,25,30])
              .addGrid(rf.maxBins, [20, 60])
```

```
            .addGrid(rf.numTrees, [5, 20,50,100])
            .build())
[In]: cv = CrossValidator(estimator=rf, estimatorParamMaps=
paramGrid, evaluator=rf_evaluator, numFolds=5)

[In]: cv_model = cv.fit(train)

[In]: best_rf_model = cv_model.bestModel
```

best_rf_model contains the best hyper-parameters to be used for training the model on this dataset.

```
[In]: model_predictions = best_rf_model.transform(test)
[In]: rf_evaluator = BinaryClassificationEvaluator(metricName='
areaUnderROC')
[In]: rf_auroc = rf_evaluator.evaluate(model_predictions)

[In]: print(rf_auroc)
[Out]:   0.7425990374615659
```

As you can see, by using the best hyperparameters for our random forest model, the AUC score has increased.

Conclusion

This chapter covered in detail the different types of supervised learning and ways to solve binary classification with multiple machine learning algorithms. How to choose the best hyperparameters for a model and using cross-validation techniques to build the best possible model on the given dataset were also explained.

CHAPTER 7

Unsupervised Machine Learning

As the name suggests, unsupervised machine learning does not include finding relationships between input and output. To be honest, there is no output that we try to predict in unsupervised learning. It is mainly used to group together the features that seem to be similar to one another in some sense. These can be the distance between those features or some sort of similarity metric. In this chapter, I will touch on some unsupervised machine learning techniques and build one of the machine learning models, using PySpark to categorize users into groups and, later, to visualize those groups as well.

Unsupervised Machine Learning Primer

As suggested, unsupervised learning does not aim to map the relationship between input and output; rather, it tries to group values that are similar to one another. There is no training that takes place on the input data. Rather, it does this by finding the underlying signals and patterns in the data, to form groups within. Unsupervised learning can be categorized further into two separate categories:

1. Clustering

2. Association rules

© Pramod Singh 2019
P. Singh, *Learn PySpark*, https://doi.org/10.1007/978-1-4842-4961-1_7

Clustering refers to finding underlying groups of data points, based on the attributes present in the data, as shown in Figure 7-1.

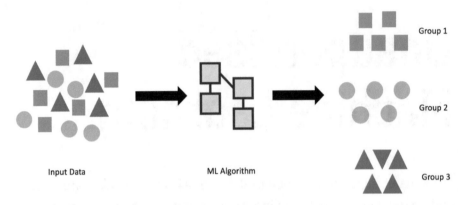

Figure 7-1. *Unsupervised machine learning*

Association signifies the probability of co-occurrence. For example, if someone bought item X, what's the probability that he or she might also buy item Y with it. In this chapter, I will focus only on clustering techniques. We can easily make use of clustering to understand the different groups in the data. For example, if we have some data about soccer players, we can easily predict who plays what position on the field—either they are forward players or defenders. Table 7-1 shows some sample data for players in a soccer tournament. The two values captured are

1. Total number of goal attempts

2. Total number of tackles made

Table 7-1. *Attempts vs. Goals*

Sr. No	No. of Goal Attempts	No. of Tackles
1	8	2
2	2	10
3	9	1
4	15	1
5	2	17
6	4	6
7	8	2
8	0	25
9	1	17
10	0	15

If we try to visualize the preceding values on a scatter plot, we get something similar to what is shown in Figure 7-2. The x axis indicates the number of tackles made by the players, and the y axis shows the number of attempts to score a goal.

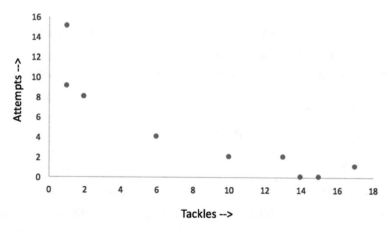

Figure 7-2. *Scatter plot of attempts vs. goals scored*

Up to this point, we don't know which players are strikers, midfielders, or defenders, but if we use some sort of clustering technique on this data, we get the groups, as shown in Figure 7-3. The figure clearly shows that there are three groups in this data. The top-left corner of the chart groups all the strikers, and the bottom-right cluster represents the defenders with the highest number of tackles and fewest attempts to score. There is also a player between these two groups who probably belongs to the midfield category, having a reasonable number of tackles and attempts to score a goal.

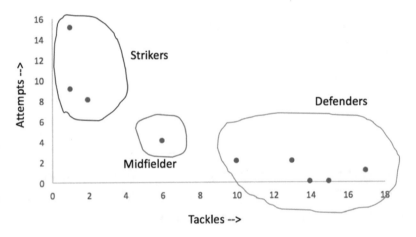

Figure 7-3. *Groups within the data*

In this example, we see how we can still predict the groups present in the data without any supervised training at all.

Clustering can have multiple applications, including the following:

1. Anomaly detection

2. Predictive maintenance

3. Customer segmentation

There are a number of clustering algorithms that can be used, depending on the available data and specific requirement. In this chapter, I am going to focus on k-means algorithms, to create clusters for sets of users who listen to different genres of music online. K-means is one of the strongest algorithms for this grouping exercise. *K* stands for the number of clusters or groups that must be formed from the data. K-means works by calculating the distance of each data point from the rest and tries to group the nearest ones, until the desired number of K values is reached. We will try to divide users into meaningful groups, using k-means algorithm, so that recommendations can be made according to their tastes in a specific musical genre.

Reviewing the Dataset

The dataset that we are going to use for this example is a sampled dataset that contains only two columns: the user id and the music category. There are close to half a million records available in this dataset.

Importing SparkSession and Creating an Object

The first step is to import SparkSession and create a Spark object, in order to use PySpark.

```
[In]: from pyspark.sql import SparkSession
[In]: spark=SparkSession.builder.appName('unsupervised_
learning').getOrCreate()
```

We will also import several other libraries, such as Pandas and NumPy, for later use.

```
[In]: import pyspark
[In]: import pandas as pd
[In]: import numpy as np
[In]: import matplotlib.pyplot as plt
[In]: from pyspark.sql.functions import *
[In]: from pyspark.sql.types import *
```

```
[In]: from pyspark.ml.clustering import KMeans
```

```
[In]: df=spark.read.csv('music_data.csv',inferSchema=True,
header=True)
[In]: df.count()
[Out]: 429023
[In]: df.printSchema()
[Out]:
```

```
root
 |-- user_id: string (nullable = true)
 |-- music category: string (nullable = true)
```

The preceding confirms that we are dealing with only two columns in the dataset. We can further explore this dataframe by doing a few aggregations. The total number of distinct music categories can be determined with the distinct function. The topmost preferred music category among users can be found by sorting the count values grouped under the music category column, as shown following:

```
[In]: df.select('music category').distinct().count()
[Out]: 21
[In]:df.groupBy('music category').count().orderBy('count',
ascending=False).show(100,False)
[Out]:
```

```
+------------------+-----+
|music category    |count|
+------------------+-----+
|Alternative Music |64227|
|Blues             |54647|
|Classical Music   |43117|
|Country Music     |39891|
|Dance Music       |35779|
|Easy Listening    |34356|
|Electronic Music  |30985|
|Rap               |29569|
|Hip Hop           |16895|
|Indie Pop         |16244|
|Motivational      |10118|
|Asian Pop         |8444 |
|Jazz              |7685 |
|Latin Music       |7302 |
|New Age           |7053 |
|Opera             |6188 |
|Hindi Music       |5689 |
|Popular           |4729 |
|Soulful           |2403 |
|Reggae            |1970 |
|Rock              |1732 |
+------------------+-----+
```

Alternative music and blues seem to be the genres most preferred by the users. We can also find the total number of unique users in the system in a manner similar to the one preceding.

```
[In]: df.select('user_id').distinct().count()
[Out]: 775
[In]:df.groupBy('user_id').count().orderBy('count',
ascending=False).show(20,False)
[Out]:
```

```
+------------------------------------------------------------------+-----+
|user_id                                                           |count|
+------------------------------------------------------------------+-----+
|628119a73b9725466e6c309f803d30cfc3b11d2a426cdd5f0d62a496e105c914|14257|
|179dd34be075e729f14dadc28a34459226c4f62797af5f186bddbb7428b8fc78|11804|
|2d765ea4ffc6554c95a1c703afcd3470bc118a65c2f0728871f4d0f40628f41b|11001|
|ca60aad85306a7fdd51a469ca6a5d27109dfd5f5781d7da985c70574d2520d23|10479|
|03bfd7e24d3a7ce9e3973f9d42ad89e4c1710b89bec59b2b08712ba73372dcbc|10075|
|7660ca8a4f5df748976aac001d5f0c5b7c8806f5c08197542d26c2eac4b20f85|9932 |
|b6eb8ef9cacc79b6779a7e065c3b0fb5174397ea983d70453bc978b6e46f95e2|7694 |
|e92f348b4a29089c9d7b82095d0bd49422153b60bbde08527015121e44699990|7210 |
|da03e0ef521fecb209f01410eefff11c115fda2d217dde51fd870761b31bb968|7182 |
|04cd1fc5c925a8a702607d6c1a047b8e0cbe170bdaed5270369d321f73254b53|6853 |
|a2374ccb47b0af60aafabeed11b979acf0d72e1afe7f96b671302aaa51592531|6730 |
|1ed20907b0eb013b6d2b8097671c5faaaae884ac9ae4b77753ee7d640036e080|6612 |
|19152e9dc36bd8fdded0be20e1a371cc5a57edd5a248973c9fa5b8ed4c8bdb0c|5378 |
|dd69f69a8df66f0d79c8a2ce41ace204dbb74e1a5a3ae36c511b8cd9ef9000a5|5255 |
|1b2dfdebfa64564c8eb7785c77b9f1cdaee6059b63b30a59c2701fee6c20407c|4928 |
|7cc577809aedea3d9423fccf6091258b9a7a0b66089d6aa86689548139c36e4b|4450 |
|0d5b7fbb167130690b8a7149d81529a2048242945a64b46c427915e8fded9e75|4371 |
|7a39568d24f3b1a953c2bc98d638cdc2b7eb02f0e6ddcee29bbc5034ec9ad76e|4324 |
|e6f90a69588f335223ac774391f7a1835d9225dc8f2090fd59265d9836783f25|4301 |
|4ab70e2a682304ce1b4b3b292b015ae343f255aca18863a4fe47a69ffd3470f8|4277 |
+------------------------------------------------------------------+-----+
only showing top 20 rows
```

```
[In]:df.groupBy('user_id').count().
orderBy('count',ascending=True).show(20,False)
[Out]:
```

```
+----------------------------------------------------------------+-----+
|user_id                                                         |count|
+----------------------------------------------------------------+-----+
|bdb448f271ec83b324acd32248f1401964cb1f5f716e94eebed508f02e889e14|1    |
|4d5920f000b000d60fdc63043886f47cc056d2835d3a73e50f389a5694b19115|1    |
|597affc56ba81274561da2b3541a8e1a30f7bbcf75cb54a9f9c1a885d88f7f7b|1    |
|ba35ab150a757e0b3abf9c5f11b7686e5394f6cdcb59d257f9fd2e1dd7cf23c2|1    |
|7ba2fabcbbd4705128a8966d4d5fcf9d3e284c77fada7752da397f479ad4a12e|1    |
|402c7aac3776a3758ca143819a0653c6cd33c6c35dd831d9ba5aad1dff43a888|1    |
|ecb27d9bf54e751018cd6c1cc5a0f2d812ff08585271acfc980bdacb742a7a02|1    |
|8aced64c96741dbeab4659f102f34c157189b3f7a63807d87bad5d8c663005cb|1    |
|a09074b27cd4ab61654c0f4e21bc5e97dc86a611743a1f8b37db3a76df8e306e|1    |
|e6e6ceb674b7ca53391ac404558a3f8d222db82c643b8c21070db4f029a50095|1    |
|8caf42bf00baea90141dec822bece26fc54bf9884c7dbf0439b445fb98b50152|1    |
|1e3e7153bf1aa5f43b1d1f27c940087587a50d40d399f4cf31b58546a39b9f26|1    |
|55efc631da61b192de874d5e2b05ea1e04ec08bd2e3009ebcd0053a84bbd4341|1    |
|760430c89527a8b503dd02bada46926938c5ad036b7ca08a21d685fe5a89f01e|1    |
|d631ffaf7f6731b146cecbf16c02fc543d02f5a8377561e5dd60a9eba603e001|1    |
|1e6fb89e6aca7641dba8bbd87e97badb8306f1ce50c02c2b0a6f36457e843221|1    |
|54f2e60bd31a775aca7fb60816c6990a415fe66694098d1ee64cc9e13f831a64|1    |
|8691b41f5ae65644c8aca0ec9d738eb3b5537b106f0fdcb3f652cc77b7ab7dd6|1    |
|205f3e9ec7c33098d85500bcf5dd0886526ee1bf1c0609ad5358994bc0a632e6|2    |
|964c0029f932feb9c7ecea499317d9e0c40020d3276e9956d71a7887321369ee|2    |
+----------------------------------------------------------------+-----+
only showing top 20 rows
```

In this dataframe, we have a mix of users. Some have listened to songs as many as 14,000 times, and others have listened only once. The next task is to reshape this dataframe, in order to use clustering on it.

Reshaping a Dataframe for Clustering

We pivot the data on user ID and music category and fill the values with the total count of songs the user has listened to. We use the crosstab function to pivot the data.

```
[In]: feature_df=df.stat.crosstab("user_id", "music category")
[In]: feature_df.printSchema()
[Out]:
```

```
root
 |-- user_id_music category: string (nullable = true)
 |-- Alternative Music: long (nullable = true)
 |-- Asian Pop: long (nullable = true)
 |-- Blues: long (nullable = true)
 |-- Classical Music: long (nullable = true)
 |-- Country Music: long (nullable = true)
 |-- Dance Music: long (nullable = true)
 |-- Easy Listening: long (nullable = true)
 |-- Electronic Music: long (nullable = true)
 |-- Hindi Music: long (nullable = true)
 |-- Hip Hop: long (nullable = true)
 |-- Indie Pop: long (nullable = true)
 |-- Jazz: long (nullable = true)
 |-- Latin Music: long (nullable = true)
 |-- Motivational: long (nullable = true)
 |-- New Age: long (nullable = true)
 |-- Opera: long (nullable = true)
 |-- Popular: long (nullable = true)
 |-- Rap: long (nullable = true)
 |-- Reggae: long (nullable = true)
 |-- Rock: long (nullable = true)
 |-- Soulful: long (nullable = true)
```

[In]: feature_df.show(3,False)
[Out]:

```
+--------------------------------------------------------------------+-----------------+---------+-----+---------------+-
-------+-----------+--------------+--------------+-----------+-------+---------+----+-----------+------------+
|user_id_music category                                              |Alternative Music|Asian Pop|Blues|Classical Music|C
ountry Music|Dance Music|Easy Listening|Electronic Music|Hindi Music|Hip Hop|Indie Pop|Jazz|Latin Music|Motivational|
New Age|Opera|Popular|Rap|Reggae|Rock|Soulful|
+--------------------------------------------------------------------+-----------------+---------+-----+---------------+-
-------+-----------+--------------+--------------+-----------+-------+---------+----+-----------+------------+
|ca3c4285512798abe7d81fa2d0588549211be9516d9b2050e5564f8e912312c7|65               |0        |7    |8              |7
|12     |7         |2             |12            |7          |0      |15       |0   |2          |0           |3
|0      |3   |0   |1   |0      |
|b457919ae168ac12ec7e0303ae0d5fe292b63c9d04eb54e039942e7f46552ec4|80               |0        |65   |135            |1
6      |55        |7             |34            |0          |12     |6        |7   |3          |2           |1
|1      |2   |0   |5   |0      |0   |0      |
|a87bb168ba6e5a8da5ecccce18c95565d843b40c1833b0f6c3089eb083f0fe3d3|0                |0        |11   |0              |9
|3      |0         |3             |3             |0          |5      |0        |0   |1          |0           |0
|4      |0   |0   |0   |5      |
+--------------------------------------------------------------------+-----------------+---------+-----+---------------+-
-------+-----------+--------------+--------------+-----------+-------+---------+----+-----------+------------+
only showing top 3 rows
```

[In]: from pyspark.ml.linalg import Vectors
[In]: from pyspark.ml.feature import VectorAssembler

Now that we have the required attributes in the new dataframe, we must assemble them, to create a single feature vector, using VectorAssembler. One key thing to remember here is that we don't use the index of the dataframe that essentially contains the user IDs for VectorAssembler.

```
[In]: print(feature_df.columns)
[Out]:
```

```
['user_id_music category', 'Alternative Music', 'Asian Pop', 'Blues', 'Classical Music', 'Country Music', 'Dance Musi
c', 'Easy Listening', 'Electronic Music', 'Hindi Music', 'Hip Hop', 'Indie Pop', 'Jazz', 'Latin Music', 'Motivationa
l', 'New Age', 'Opera', 'Popular', 'Rap', 'Reggae', 'Rock', 'Soulful']
```

```
[In]: feat_cols=[col for col in feature_df.columns if col !=
'user_id_music category']
[In]: print(feat_cols)
[Out]:
```

```
['Alternative Music', 'Asian Pop', 'Blues', 'Classical Music', 'Country Music', 'Dance Music', 'Easy Listening', 'Ele
ctronic Music', 'Hindi Music', 'Hip Hop', 'Indie Pop', 'Jazz', 'Latin Music', 'Motivational', 'New Age', 'Opera', 'Po
pular', 'Rap', 'Reggae', 'Rock', 'Soulful']
```

```
[In]: vec_assembler = VectorAssembler(inputCols = feat_cols,
outputCol='features')
[In]: final_data = vec_assembler.transform(feature_df)

[In]: final_data.printSchema()
[Out]:
```

```
root
 |-- user_id_music category: string (nullable = true)
 |-- Alternative Music: long (nullable = true)
 |-- Asian Pop: long (nullable = true)
 |-- Blues: long (nullable = true)
 |-- Classical Music: long (nullable = true)
 |-- Country Music: long (nullable = true)
 |-- Dance Music: long (nullable = true)
 |-- Easy Listening: long (nullable = true)
 |-- Electronic Music: long (nullable = true)
 |-- Hindi Music: long (nullable = true)
 |-- Hip Hop: long (nullable = true)
 |-- Indie Pop: long (nullable = true)
 |-- Jazz: long (nullable = true)
 |-- Latin Music: long (nullable = true)
 |-- Motivational: long (nullable = true)
 |-- New Age: long (nullable = true)
 |-- Opera: long (nullable = true)
 |-- Popular: long (nullable = true)
 |-- Rap: long (nullable = true)
 |-- Reggae: long (nullable = true)
 |-- Rock: long (nullable = true)
 |-- Soulful: long (nullable = true)
 |-- features: vector (nullable = true)
```

Note One thing to remember is that clustering can be sensitive to the scale of the data, for the simple reason that it uses a distance metric to compare the similarity between two values (multidimensionality). Therefore, it's always a good idea to scale down the data before applying clustering.

Next, we import StandardScaler in the Spark library and apply it on the feature vectors.

```
[In]: from pyspark.ml.feature import StandardScaler
[In]:scaler = StandardScaler(inputCol="features",
outputCol="scaledFeatures", withStd=True, withMean=False)
[In]: scalerModel = scaler.fit(final_data)
```

```
[In]: cluster_final_data.columns
[Out]:
```

```
['user_id_music category',
 'Alternative Music',
 'Asian Pop',
 'Blues',
 'Classical Music',
 'Country Music',
 'Dance Music',
 'Easy Listening',
 'Electronic Music',
 'Hindi Music',
 'Hip Hop',
 'Indie Pop',
 'Jazz',
 'Latin Music',
 'Motivational',
 'New Age',
 'Opera',
 'Popular',
 'Rap',
 'Reggae',
 'Rock',
 'Soulful',
 'features',
 'scaledFeatures']
```

The next step is to actually build clusters, using a k-means clustering algorithm.

Building Clusters with K-Means

One of the key questions to ask before a clustering exercise is: "How many clusters should be formed from the data?" One way to approach this issue is with the Elbow method, by which we try to plot the total sum of squared errors within a cluster against the number of clusters. This gives us a sense of what's a good number of groups that can be formed from the given data. Basically, we apply k-means for a set consisting of a pre-decided number of clusters (2–10) and visualize the errors against it. Wherever there is an elbow kind of shape forming on the chart, that is a good number to pick for k.

```
[In]: errors=[]
for k in range(2,10):
    kmeans = KMeans(featuresCol='scaledFeatures',k=k)
    model = kmeans.fit(cluster_final_data)
    wssse = model.computeCost(cluster_final_data)
    errors.append(wssse)
    print("With K={}".format(k))
    print("Within Set Sum of Squared Errors = " + str(wssse))
    print('--'*30)
```

[Out]:

```
With K=2
Within Set Sum of Squared Errors = 13324.475648979027
------------------------------------------------------------
With K=3
Within Set Sum of Squared Errors = 12699.899722749344
------------------------------------------------------------
With K=4
Within Set Sum of Squared Errors = 11263.261281238181
------------------------------------------------------------
With K=5
Within Set Sum of Squared Errors = 11608.125811487394
------------------------------------------------------------
With K=6
Within Set Sum of Squared Errors = 10479.14665326082
------------------------------------------------------------
With K=7
Within Set Sum of Squared Errors = 10377.739238575698
------------------------------------------------------------
With K=8
Within Set Sum of Squared Errors = 8884.406507460375
------------------------------------------------------------
With K=9
Within Set Sum of Squared Errors = 8743.346814279732
------------------------------------------------------------
```

```
[In]: cluster_number = range(2,10)
[In]: plt.scatter(cluster_number,errors)
[In]: plt.xlabel('clusters')
```

```
[In]: plt.ylabel('WSSE')
[In]: plt.show()
[Out]:
```

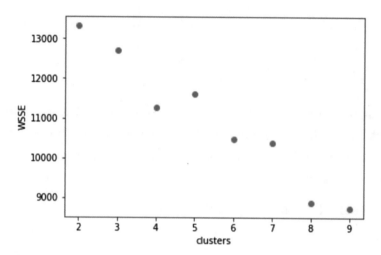

Figure 7-4. *Elbow chart*

In our case, there are multiple elbows being formed in the chart. Therefore, we can choose either 4, 6, or 8 as the value of k. One thing to remember is that there is no correct answer in clustering. You can have different clusters from the same data, based on techniques being used, as the initiation points differ for every technique. Let's go ahead and chose the value of k as 6.

```
[In]: kmeans6 = KMeans(featuresCol='scaledFeatures',k=6)
[In]: model_k6 = kmeans6.fit(cluster_final_data)
[In]: model_k6.transform(cluster_final_data).
groupBy('prediction').count().show()
```

[Out]:

```
+----------+-----+
|prediction|count|
+----------+-----+
|         1|    6|
|         3|   14|
|         5|    1|
|         4|    3|
|         2|    2|
|         0|  749|
+----------+-----+
```

From our data, we see that there is one cluster in which the majority of users belong. Apart from that, there are five more clusters with lesser values. Now, we add this prediction to the existing dataframe, to be able to visualize the clusters.

[In]: model_k6.transform(cluster_final_data).show()
[Out]:

```
+--------------------+-----------------+---------+-----+---------------+--------------+-----------+--------------+----------+--------+------------+-------------+----------+---------+-------+------+----+------+-----+--------------------+--------------------+----------+
|user_id_music category|Alternative Music|Asian Pop|Blues|Classical Music|Country Music|Dance Music|Easy Listening|Electronic Music|Hindi Music|Hip Hop|Indie Pop|Jazz|Latin Music|Motivational|New Age|Opera|Popular|Rap|Reggae|Rock|Soulful|            features|       scaledFeatures|prediction|
+--------------------+-----------------+---------+-----+---------------+--------------+-----------+--------------+----------+--------+------------+-------------+----------+---------+-------+------+----+------+-----+--------------------+--------------------+----------+
|  ca3c4285512798abe...|               65|        0|    7|              8|            7|         12|             7|         12|       7|           0|           15|         0|        2|      0|     3|   0|     3|    0|     1|   0|[65.0,0...|0,7.0,8.0...|[0.23836018560172...|          0|
|  b457919ae168ac12e...|               80|        0|   65|            135|           16|         55|             7|         34|       0|           12|            6|         7|        3|      0|     2|   1|     2|    0|     5|   0|[80.0,0...|0,65.0,13...|[0.29336638227904...|          0|
|  a87bb168ba6e5a8da...|                0|        0|   11|              0|            9|          3|             0|          3|       3|           0|            5|         0|        0|      1|     0|   0|     4|    0|     0|   5|[21,[2,4,...|5,7,8,10...|(21,[2,4,5,7,8,10...|          3|
|  cf60c9b1a0444261c...|              119|      268|  340|             63|           85|         48|           498|         82|      0|           34|            1|        14|        0|     0|    14|   3|    57|    0|    67|   3|[119.0,2...|68.0,340...|[0.43638249364008...|          1|
|  8ab21c1b361d5c4aa...|               23|        7|   81|             22|           30|        197|            11|          8|       0|            7|            6|         0|        0|     0|    12|   7|     2|    9|     7|   0|[23.0,7...|0,81.0,22...|[0.08434283490522...|          0|
|  34e22754ffec4a0d4...|              192|        1|  158|             40|           83|        110|           126|         39|      1|           33|           68|         1|        0|      0|     4|   2|     0|   11|    13|   0|[192.0,...|1.0,158.0,...|[0.70407931746971...|          7|
|  ad3ea6999df3120cc...|                5|        0|    1|              0|            9|        254|            15|          1|      2|            0|            4|         0|        1|      0|     0|   0|     4|    0|     1|   0|[21,[0,2,...|4,5,6,7,...|(21,[0,2,4,5,6,7,...|          0|
|  3d2ea50eeeb98d4a7...|               31|        2|   17|             22|           21|         66|            90|          8|      2|           26|           42|        10|        0|      4|     0|   6|    12|    0|    14|   0|[31.0,2....|0,17.0,22...|[0.11367947313313...|          9|
|  334533d21eeed37dd...|                7|        0|    5|              0|            4|          4|             1|          0|      2|            4|            3|         0|        0|      0|     0|   2|     3|    0|     0|   0|[21,[0,2,...|4,5,6,8,...|(21,[0,2,4,5,6,8,...|          2|
|  bad9c123790cfd0b1...|               19|        3|    7|             19|           13|         33|             7|          4|     49|           20|            5|         0|        0|     0|     0|  99|     3|    0|    15|   0|[19.0,3....|0,7.0,19....|[0.06967451579127...|          0|
+--------------------+-----------------+---------+-----+---------------+--------------+-----------+--------------+----------+--------+------------+-------------+----------+---------+-------+------+----+------+-----+--------------------+--------------------+----------+
only showing top 10 rows
```

```
[In]: cluser_prediction=model_k6.transform(cluster_final_data)
[In]: cluser_prediction.printSchema()
[Out]:
```

```
root
 |-- user_id_music category: string (nullable = true)
 |-- Alternative Music: long (nullable = true)
 |-- Asian Pop: long (nullable = true)
 |-- Blues: long (nullable = true)
 |-- Classical Music: long (nullable = true)
 |-- Country Music: long (nullable = true)
 |-- Dance Music: long (nullable = true)
 |-- Easy Listening: long (nullable = true)
 |-- Electronic Music: long (nullable = true)
 |-- Hindi Music: long (nullable = true)
 |-- Hip Hop: long (nullable = true)
 |-- Indie Pop: long (nullable = true)
 |-- Jazz: long (nullable = true)
 |-- Latin Music: long (nullable = true)
 |-- Motivational: long (nullable = true)
 |-- New Age: long (nullable = true)
 |-- Opera: long (nullable = true)
 |-- Popular: long (nullable = true)
 |-- Rap: long (nullable = true)
 |-- Reggae: long (nullable = true)
 |-- Rock: long (nullable = true)
 |-- Soulful: long (nullable = true)
 |-- features: vector (nullable = true)
 |-- scaledFeatures: vector (nullable = true)
 |-- prediction: integer (nullable = false)
```

Because we're dealing with multiple dimensions, it will become difficult to visualize the data with cluster numbers. Therefore, we reduce the total number of dimensions, using the PCA (principal component analysis) technique. In previous chapters, you have already seen how to use PCA. We now reduce the original number of features from 21 to just 3, using PCA, as shown following.

```
[In]: from pyspark.ml.feature import PCA
[In]: from pyspark.ml.linalg import Vectors
```

```
[In]: pca = PCA(k=3, inputCol="scaledFeatures", outputCol="pca_
features")
[In]: pca_model = pca.fit(cluser_prediction)

[In]: result = pca_model.transform(cluser_prediction).
select('user_id_music category',"pca_features",'prediction')
[In]: result.show(truncate=False)
[Out]:
```

```
+--------------------+----------+                                              +------------------------------------------------
|user_id_music category                                                        |pca_features
|prediction|
+--------------------+----------+                                              +------------------------------------------------
|ca3c4285512798abe7d81fa2d0588549211be9516d9b2050e5564f8e912312c7|[-0.2250735026516284,0.1508289363663226,-0.13922132
997808534]          |0         |
|b457919ae168ac12ec7e0303ae0d5fe292b63c9d04eb54e039942e7f46552ec4|[-0.5273003162834193,0.10206886950980007,-0.0149713
60310926437]        |0         |
|a87bb168ba6e5a8da5eccce18c95565d843b40c1833b0f6c3089eb083f0fe3d3|[-0.23968261020295822,0.2801153618243124,-0.1738219
443638187]          |0         |
|cf60c9b1a0444261c3a4687677f5336686b96756df5e33c11f6016f16fb38372|[-3.5120515919930178,0.32729433189573715,1.10173026
17933256]           |0         |
|8ab21c1b361d5c4aa1ed01893305febf1c6659bd342217eece1225d789ba57da|[-0.6796385576703683,0.39608053546483746,0.11474085
857053855]          |0         |
|34e22754ffec4a0d434edfa0f3b10c0e1ab14c632cee37514076d211cf1e7017|[-2.0964050329364166,0.9121320035474169,0.991261384
6332612]            |0         |
|ad3ea6999df3120ccb557d34cf75e17dfdd3febe445e9e1bc36514642fcd3bdb|[-0.2944449488892162,0.383978047265129,0.0775875209
5209272]            |0         |
|3d2ea50eeeb98d4a79259ee057955696ec9dc46e1b72e8d8eff099feb3362e1f|[-0.9356027216181408,0.595824057721615,-0.175241540
04372516]           |0         |
|334533d21eeed37dd2b686022857f9f42eef9fa6192a8653040af844be8d15ef|[-0.13886158277802202,0.15784222493276218,-0.112606
24144367197]        |0         |
|bad9c123790cfd0b1106190de5623afbb70de0e71f5d48dc05ee6d743a2f50a8|[-0.5824585609741292,0.6698195803354224,-0.86840549
60184615]           |0         |
|dd8b41323bbef6d4093879c63966b025d9dd3dbdab1b372ec3c09ee9e40e277c|[-0.196130385293706,0.12812984441339165,-0.09549024
406090195]          |0         |
|ac40c16b3c6248cebdfb8cb9e2dee4575862ef1223ece1b393b45b6b2561e8fc|[-0.28548897643969706,0.1507484298534756,-0.2556035
9089598994]         |0         |
|57c20a25da427c79b41690be1b895110bbae9fb583a32a2d27c310e9caa1a395|[-0.11298754622944404,-0.02683169465747653,0.012421
222276786992]|0         |
|b5773ac4aa0d079266ecc3f02d0a16a0f7fd0789b7c0553782dd0588d57298b4|[-0.4918021450678279,0.3841542970784488,-0.10281962
873038579]          |0         |
|793c0773daa478bf340a4c6f63fad3beedef69aaa0a8f503995450c8a75cd2bd|[-0.05930380568903517,0.03756596949668735,0.0826018
1325447052]         |0         |
|4c52a409ecbf4c78bf6aa73611615d7c68090162250aeed3118d57177e291e63|[-1.4580702292222283,1.8999456814099018,-3.77150115
79509974]           |0         |
|fadebf953475576be81797433f0d28278858f93eee0cefcc95b5b03dd787c327|[-0.7083739895738098,0.21507912946609467,0.35753821
912208134]          |0         |
|86f98ea95c8b3333b947bf9a1a305150633df37e6aac807cddcbbe3463bfd53d|[-2.4658708965532194,0.6606526791365808,0.322308658
42212864]           |0         |
|1a9da82f18b16577b872a71cefdeae63a7725eac5547820182ba82d8845b02ee|[-4.31655963988883,2.1119157004751234,1.54934869211
12498]              |0         |
|9e528e07d7deddba5dcafd78c60059c7a91a7233350b2dcad109ae3b413d6aaf|[-0.8673164731150801,0.025748509689555762,0.2245798
4151934027]         |0         |
+--------------------+----------+                                              +------------------------------------------------
only showing top 20 rows
```

Finally, we convert the PCA dataframe to a Pandas dataframe, in addition to creating separate columns (x, y, z) from the PCA feature column, by making individual columns from the list.

```
[In]: clusters = result.toPandas().set_index('user_id_music
category')
[In]: clusters.head(10)
[Out]:
```

	pca_features	prediction
user_id_music category		
ca3c4285512798abe7d81fa2d0588549211be9516d9b2050e5564f8e912312c7	[-0.225073502652, 0.150828936366, -0.139221329...	0
b457919ae168ac12ec7e0303ae0d5fe292b63c9d04eb54e039942e7f46552ec4	[-0.527300316283, 0.10206886951, -0.0149713603...	0
a87bb168ba6e5a8da5eccce18c95565d843b40c1833b0f6c3089eb083f0fe3d3	[-0.239682610203, 0.280115361824, -0.173821944...	0
cf60c9b1a0444261c3a4687677f5336686b96756df5e33c11f6016f16fb38372	[-3.51205159199, 0.327294331896, 1.10173026179]	0
8ab21c1b361d5c4aa1ed01893305febf1c6659bd342217eece1225d789ba57da	[-0.67963855767, 0.396080535465, 0.114740858571]	0
34e22754ffec4a0d434edfa0f3b10c0e1ab14c632cee37514076d211cf1e7017	[-2.09640503294, 0.912132003547, 0.991261384633]	0
ad3ea6999df3120ccb557d34cf75e17dfdd3febe445e9e1bc36514642fcd3bdb	[-0.294444948889, 0.383978047265, 0.0775875209...	0
3d2ea50eeeb98d4a79259ee057955696ec9dc46e1b72e8d8eff099feb3362e1f	[-0.935602721618, 0.595824057722, -0.175241540...	0
334533d21eeed37dd2b686022857f9f42eef9fa6192a8653040af844be8d15ef	[-0.138861582778, 0.157842224933, -0.112606241...	0
bad9c123790cfd0b1106190de5623afbb70de0e71f5d48dc05ee6d743a2f50a8	[-0.582458560974, 0.669819580335, -0.868405496...	0

```
[In]: clusters[['x','y','z']]=pd.DataFrame(clusters.pca_
features.values.tolist(), index= clusters.index)
[In]: del clusters['pca_features']
[In]: clusters.head(10)
[Out]:
```

	prediction	x	y	z
user_id_music category				
ca3c4285512798abe7d81fa2d0588549211be9516d9b2050e5564f8e912312c7	0	-0.225074	0.150829	-0.139221
b457919ae168ac12ec7e0303ae0d5fe292b63c9d04eb54e039942e7f46552ec4	0	-0.527300	0.102069	-0.014971
a87bb168ba6e5a8da5eccce18c95565d843b40c1833b0f6c3089eb083f0fe3d3	0	-0.239683	0.280115	-0.173822
cf60c9b1a0444261c3a4687677f5336686b96756df5e33c11f6016f16fb38372	0	-3.512052	0.327294	1.101730
8ab21c1b361d5c4aa1ed01893305febf1c6659bd342217eece1225d789ba57da	0	-0.679639	0.396081	0.114741
34e22754ffec4a0d434edfa0f3b10c0e1ab14c632cee37514076d211cf1e7017	0	-2.096405	0.912132	0.991261
ad3ea6999df3120ccb557d34cf75e17dfdd3febe445e9e1bc36514642fcd3bdb	0	-0.294445	0.383978	0.077588
3d2ea50eeeb98d4a79259ee057955696ec9dc46e1b72e8d8eff099feb3362e1f	0	-0.935603	0.595824	-0.175242
334533d21eeed37dd2b686022857f9f42eef9fa6192a8653040af844be8d15ef	0	-0.138862	0.157842	-0.112606
bad9c123790cfd0b1106190de5623afbb70de0e71f5d48dc05ee6d743a2f50a8	0	-0.582459	0.669820	-0.868405

Now that we have three-dimensional (3D) data representing the original features, and the cluster prediction by k-means, we can use 3D plotting techniques, to visualize these clusters on a 3D plot.

```
[In]: from mpl_toolkits.mplot3d import Axes3D
[In]: cluster_vis= plt.figure(figsize=(10,10)).
gca(projection='3d')
[In]: cluster_vis.scatter(clusters.x, clusters.y, clusters.z,
c=clusters.prediction)
[In]: cluster_vis.set_xlabel('x')
[In]: cluster_vis.set_ylabel('y')
[In]: cluster_vis.set_zlabel('z')
[In]: plt.show()
[Out]:
```

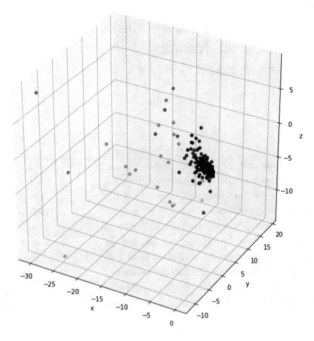

Figure 7-5. *Clusters Visualization*

One of the follow-up activities after this clustering exercise can be to assign different personas to the groups, based on the attribute values. This can be a manual activity and can take some time to come up with meaningful personas. Once created, businesses can use these in numerous ways to market and target specific users.

Conclusion

In this chapter, you learned the difference between supervised and unsupervised machine learning techniques. You also saw how to build clusters from raw data, using k-means clustering. The method to find the optimal value of k and how to visualize the final clusters formed by the k-means algorithm were then explained.

CHAPTER 8

Deep Learning Using PySpark

Deep learning has been in the limelight for quite a few years and is making leaps and bounds in terms of solving various business challenges. From image language translation to self-driving cars, deep learning has become an important component in the larger scheme of things. There is no denying the fact that lots of companies today are betting heavily on deep learning, as a majority of their applications run using deep learning in the back end. For example, Google's Gmail, YouTube, Search, Maps, and Assistance all use deep learning in some form or other. The reason is deep learning's incredible ability to provide far better results, compared to some other machine learning algorithms.

This chapter is divided into three parts. The first focuses on understanding the fundamentals and underlying operating principles of deep learning. The second part covers the training process of the deep learning model. Finally, in the third and final part, you will see how to build a multilayer perceptron, using Spark.

Deep Learning Fundamentals

Before even getting into deep learning, we must understand what neural networks are, as deep learning is a sort of extension of neural networks. Neural networks are not new; in fact, they go way back to the 1950s, when

© Pramod Singh 2019
P. Singh, *Learn PySpark*, https://doi.org/10.1007/978-1-4842-4961-1_8

researchers began working on them. Unfortunately, they hit a major roadblock, owing to limited computation capabilities at the time. In the recent past, neural networks have gained in popularity, due to major advancements in processing power and access to big data. The availability of super-powerful processing devices, such as GPUs and TPUs, has made it possible to run huge neural networks with better performance. From the data aspect, the availability of labeled data over the last few years has also helped immensely. More than any of the preceding reasons, it's the unique ability of deep learning models to offer significant performance over other machine learning algorithms that has made deep learning the preferred, mainstream approach. Figure 8-1 shows the evolutionary time line of artificial intelligence, machine learning, and deep learning.

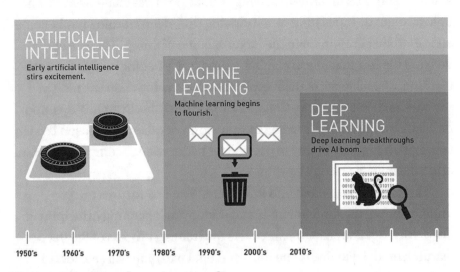

Figure 8-1. *Deep learning time line*

In supervised learning settings, there is specific input and corresponding output. The objective of machine learning algorithms is to use this data and approximate the relationship between input and output variables. In some cases, this relationship can be quite evident and easy to capture, but in realistic scenarios, the relationship between input and output variables is very complex and nonlinear in nature. To give an example, a self-driving car, the input variables can be as follows:

- Terrain

- Distance from nearest object

- Traffic lights

- Signboards

The output must be either turn or drive fast or slowly or apply brakes, etc. As you might imagine, the relationship between input variables and output variables is fairly complex in nature. Therefore, the traditional machine learning algorithm finds it hard to map this kind of relationship. However, deep learning outperforms other machine learning algorithms in such situations, as it is able to learn nonlinear features as well.

Human Brain Neuron vs. Artificial Neuron

As mentioned, deep learning is an extension of neural networks only and is also known as deep neural networks. Neural networks are a little different than other machine learning algorithms. Neural networks are loosely inspired by neurons in the human brain. Neural networks are made up of artificial neurons. Although I don't claim to be an expert on neuroscience or functioning of the brain, let me try to give you a high-level overview of how the human brain functions. You might already be aware of the fact that the human brain is made up of billions of neurons, with an incredible number of connections between them. Each neuron is connected by multiple other neurons in some way and repeatedly exchanges information (signals). Each activity that we undertake physically or mentally fires up a certain set of neurons in our brain. Every neuron is made up of three basic components:

- Dendrites

- Cell Body

- Terminals

As you can see in Figure 8-2, in a human brain neuron, the dendrites are responsible for receiving signals from other neurons. They act as receivers of the particular neuron and pass information to a cell body, where the specific information is processed. Now, based on the level of information, it either activates (fires up) or doesn't trigger. This activity depends on a particular threshold value of the neuron. If the incoming signal value is below that threshold, it will not fire; otherwise, it activates. Finally, the third component is terminals, which are connected to the dendrites of other neurons. Terminals are responsible for passing on the output of a particular neuron to other relevant connectors.

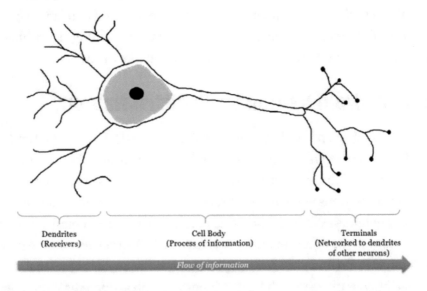

Figure 8-2. *Neuron of the human brain*

Artificial neurons, on the other hand, consist mainly of two parts: one is summation, and the other is activation, as shown in Figure 8-3. This is also known as a perceptron. Summation refers to adding all the input signals, and activation refers to whether the neuron will trigger, based on the threshold value.

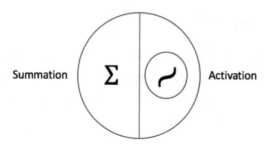

Figure 8-3. *Parts of an artificial neuron*

Let's say we have two binary inputs (X1, X2) and weights of their respective connections (W1, W2), as shown in Figure 8-4. The weights can be considered similar to coefficients of input variables in traditional machine learning. These weights indicate how important the particular input feature in the model is. The summation function calculates the total sum of the input.

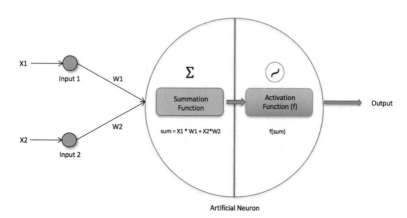

Figure 8-4. *Inputs and weights in an artificial neuron*

The activation function then uses this total summated value and gives a certain output. Activation is sort of a decision-making function. Based on the type of activation function used, it gives an output accordingly. There are different types of activation functions that can be used in a neural network layer.

Activation Functions

Activation functions play a critical role in neural networks, as the output varies, based on the type of activation function used. There are typically three main activation functions that are widely used: sigmoid, hyperbolic tangent, and rectified linear unit.

Sigmoid

This activation function ensures that the output is always between 0 and 1, irrespective of the input, as shown in Figure 8-5. That's why it is also used in logistic regression, to predict the probability of an event.

$$f(x) = \frac{1}{1+e^{-x}}$$

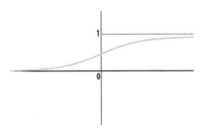

Figure 8-5. *Sigmoid activation function*

Hyperbolic Tangent

Hyperbolic tangent activation (tanh) ensures that the output value remains between -1 to 1, regardless of the input, as shown in Figure 8-6. Following is the tanh formula:

$$f(x) = \frac{e^{2x}-1}{e^{2x}+1}$$

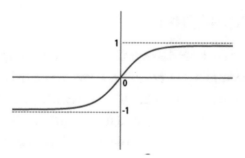

Figure 8-6. *Tanh activation function*

Rectified Linear Unit

Rectified linear units (ReLUs) have been increasingly popular over the last few years and have become the default activation function. A ReLU is very powerful, as it produces values between 0 and ∞. If the input is 0 or less than 0, the output is always going to be 0, but for anything more than 0, the output is similar to the input, as shown in Figure 8-7. The formula for a ReLU is

$$f(x) = \max(0,x)$$

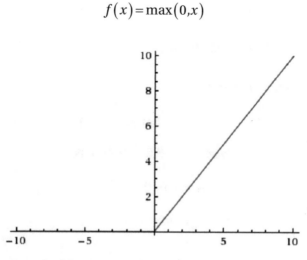

Figure 8-7. *Rectified linear unit*

Neuron Computation

Now that we have a basic understanding of different activation functions, let's consider an example, to understand how the actual output is calculated inside a neuron. Let's say we have two inputs, X1 and X2, with values of 0.2 and 0.7, respectively, and the weights are 0.05 and 0.03. The summation function calculates the total sum of incoming input signals, as shown in Figures 8-8 and 8-9.

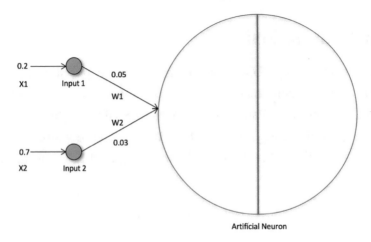

Artificial Neuron

Figure 8-8. *Neuron computation*

The summation is as follows:

$$sum = X1*W1 + X2*W2$$

$$sum = 0.2*0.05 + 0.7*0.03$$

$$sum = 0.01 + 0.021$$

$$sum = 0.031$$

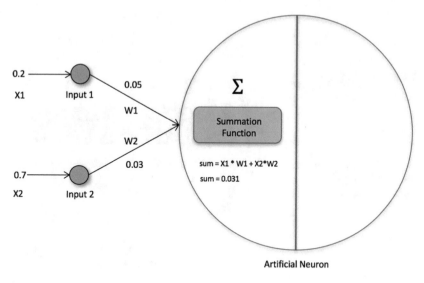

Figure 8-9. *Summation*

The next step is to pass this sum through an activation function. Let's consider using a sigmoid function, which returns values between 0 and 1, irrespective of the input. The sigmoid function will calculate the value, as follows:

$$f(x) = \frac{1}{\left(1 + e^{-x}\right)}$$

$$f(sum) = \frac{1}{\left(1 + e^{-sum}\right)}$$

$$f(0.031) = \frac{1}{\left(1 + e^{-0.031}\right)}$$

$$f(0.031) = 0.5077$$

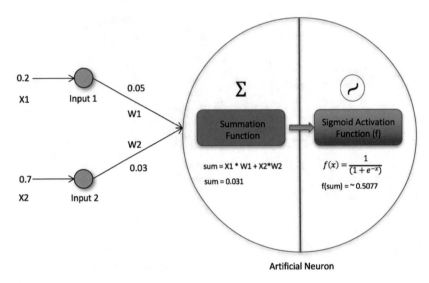

Artificial Neuron

Figure 8-10. *Neuron Activation*

So, the output of this single neuron is equal to 0.5077. Now that we know how a single neuron operates, let's quickly go over how multiple connected neurons work together to calculate the output.

Training Process: Neural Network

When we combine multiple neurons, we end up with a neural network. Most simple and basic neural networks can be built using just input and output neurons, as shown in Figure 8-11.

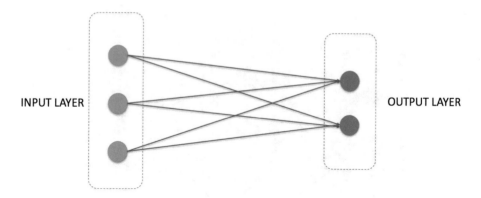

Figure 8-11. *Basic neural network*

The challenge with using a neural network such as this is that it can only learn linear relationships and cannot perform well in cases in which the relationship between input and output is nonlinear. As we have seen, in real-world scenarios, the relationship is hardly simple and linear. Therefore, we must introduce an additional layer of neurons between the input and output layer, in order to increase the network's capacity to learn different kinds of nonlinear relationships. This additional layer of neurons is known as a hidden layer, as shown in Figure 8-12. It is responsible for introducing nonlinearities into the learning process of the network. Neural networks are also known as universal approximators, because they have the ability to approximate any relationship between input and output variables, no matter how complex and nonlinear in nature. A lot depends on the number of hidden layers in the networks and the total number of neurons in each hidden layer. Given sufficient numbers of hidden layers, a network can perform brilliantly at mapping this relationship.

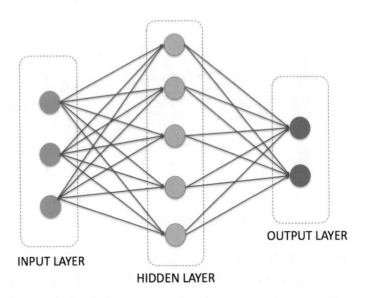

Figure 8-12. *Multiple layer neural network*

A neural network is all about various connections (red lines) and different weights associated with these connections. The training of neural networks primarily includes adjusting these weights in such a way that the model can make predictions with a higher degree of accuracy. To understand how neural networks are trained, let's break down the steps of network training.

> Step 1. Take the input values and calculate the output values that are passed to hidden neurons, as shown in Figure 8-13. The weights used for the first iteration of sum calculation are generated randomly.

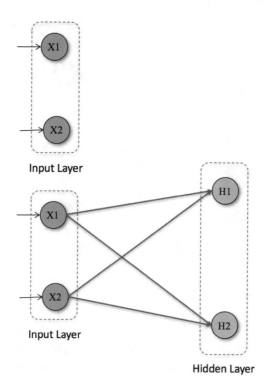

Figure 8-13. *Neural network training process*

An additional component that is passed is the bias neuron input, as shown in Figure 8-14. This is mainly used when you want to have some non-zero output for even the zero input values.

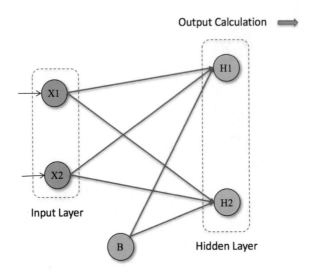

Figure 8-14. *Bias component*

Step 2. The hidden layer neurons now go through the same process to calculate the output, using the inputs from the previous layer (input layer). This hidden layer output acts as an input for the final output neuron (red) calculation, as shown in Figure 8-15.

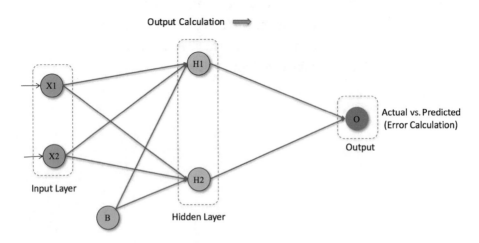

Figure 8-15. *Output calculation*

Step 3. Once we have the final output, it is compared with the actual output, and the error is backpropagated to the network, to adjust the weights of the connections so as to reduce the overall error on the training data, as shown in Figure 8-16. This process is known as backpropagation.

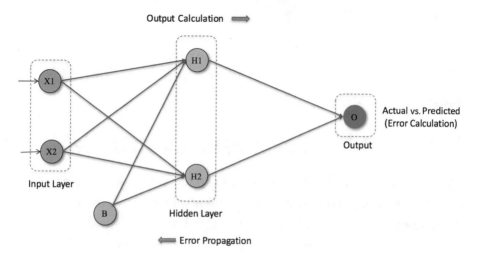

Figure 8-16. *Backpropagation*

Step 4. Weights of the connections are readjusted according to the output, to minimize the overall errors made by the network, to the point that there is no further reduction of error on the training data.

Step 5. Now that we have the final version of the weights, a new output value is calculated, based on updated weights, by the network, as shown in Figure 8-17.

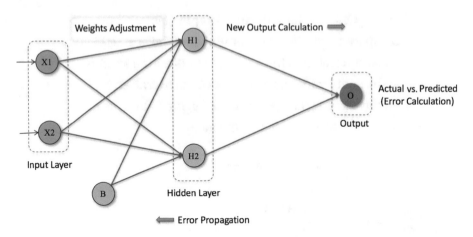

Figure 8-17. *Final output*

Building a Multilayer Perceptron Model

When it comes to using Deep Learning in Spark, there are multiple options.Depending on the exact requirement and available infrastructure, the relevant approach can be used. On a high level,there are close to 4 important deep learning libraries which can be used with Spark.

1. Spark's MLlib

2. TensorflowOnSpark

3. Deep Learning Pipelines

4. DeepLearning4J

For simplicity, we will build a multilayer perceptron, using Spark. The dataset that we are going to use for this exercise contains close to 75k records, with some sample customer journey data on a retail web site. There are 16 input features to predict whether the visitor is likely to convert. We have a balanced target class in this dataset. We will use `MultilayerPerceptronClassifier` from Spark's Machine Learning library. We start by importing a few important dependencies.

```
[In]: from pyspark.sql import SparkSession
[In]: spark = SparkSession.builder.appName('deep_learning').
getOrCreate()
[In]: import os
[In]: import numpy as np
[In]: import pandas as pd
[In]: from pyspark.sql.types import *
```

Now we load the dataset into Spark, for feature engineering and model training. As mentioned, there are 16 input features and 1 output column ('Orders_Normalized').

```
[In]: data = spark.read.csv('dl_data.csv', header=True,
inferSchema=True)
[In]: data.printSchema()
[Out]:
```

```
root
 |-- Visit_Number_Bucket: string (nullable = true)
 |-- Page_Views_Normalized: double (nullable = true)
 |-- Orders_Normalized: integer (nullable = true)
 |-- Internal_Search_Successful_Normalized: double (nullable = true)
 |-- Internal_Search_Null_Normalized: double (nullable = true)
 |-- Email_Signup_Normalized: double (nullable = true)
 |-- Total_Seconds_Spent_Normalized: double (nullable = true)
 |-- Store_Locator_Search_Normalized: double (nullable = true)
 |-- Mapped_Last_Touch_Channel: string (nullable = true)
 |-- Mapped_Mobile_Device_Type: string (nullable = true)
 |-- Mapped_Browser_Type: string (nullable = true)
 |-- Mapped_Entry_Pages: string (nullable = true)
 |-- Mapped_Site_Section: string (nullable = true)
 |-- Mapped_Promo_Code: string (nullable = true)
 |-- Maped_Product_Name: string (nullable = true)
 |-- Mapped_Search_Term: string (nullable = true)
 |-- Mapped_Product_Collection: string (nullable = true)
```

We change the name of the label column from `'Orders_Normalized'` to `'label'`, to be able to train the model.

```
[In]: data = data.withColumnRenamed('Orders_Normalized',
'label')
[In]: data.printSchema()
[Out]:
```

```
root
 |-- Visit_Number_Bucket: string (nullable = true)
 |-- Page_Views_Normalized: double (nullable = true)
 |-- label: integer (nullable = true)
 |-- Internal_Search_Successful_Normalized: double (nullable = true)
 |-- Internal_Search_Null_Normalized: double (nullable = true)
 |-- Email_Signup_Normalized: double (nullable = true)
 |-- Total_Seconds_Spent_Normalized: double (nullable = true)
 |-- Store_Locator_Search_Normalized: double (nullable = true)
 |-- Mapped_Last_Touch_Channel: string (nullable = true)
 |-- Mapped_Mobile_Device_Type: string (nullable = true)
 |-- Mapped_Browser_Type: string (nullable = true)
 |-- Mapped_Entry_Pages: string (nullable = true)
 |-- Mapped_Site_Section: string (nullable = true)
 |-- Mapped_Promo_Code: string (nullable = true)
 |-- Maped_Product_Name: string (nullable = true)
 |-- Mapped_Search_Term: string (nullable = true)
 |-- Mapped_Product_Collection: string (nullable = true)
```

Because we are dealing with both numerical and categorical columns, we must write a pipeline to create features combining both for model training. Therefore, we import `Pipeline`, `VectorAssembler`, and `OneHotEncoder`, to create feature vectors. We will also import `MultiClassificationEvaluator` and `MultilayerPerceptron`, to check the performance of the model.

```
[In]: from pyspark.ml.feature import OneHotEncoder,
VectorAssembler, StringIndexer
[In]: from pyspark.ml import Pipeline
[In]: from pyspark.sql.functions import udf, StringType
```

```
[In]: from pyspark.ml.evaluation import
MulticlassClassificationEvaluator
[In]: from pyspark.ml.classification import
MultilayerPerceptronClassifier
```

We now split the data into train, test, and validation sets, for training of the model.

```
[In]: train, validation, test  = data.randomSplit([0.7, 0.2,
0.1], 1234)
```

We create separate lists of categorical columns and numeric columns, based on datatypes.

```
[In]: categorical_columns = [item[0] for item in data.dtypes if
item[1].startswith('string')]
[In]: numeric_columns = [item[0] for item in data.dtypes if
item[1].startswith('double')]
[In]: indexers = [StringIndexer(inputCol=column,
outputCol='{0}_index'.format(column)) for column in
categorical_columns]
```

We now create consolidated feature vectors, using VectorAssembler.

```
[In]: featuresCreator = VectorAssembler(inputCols=[indexer.
getOutputCol() for indexer in indexers] + numeric_columns,
outputCol="features")
[In]: layers = [len(featuresCreator.getInputCols()), 4, 2, 2]
```

The next step is to build the MultilayerPerceptron model. One can play around with different hyperparameters, such as number of layers and maxiters, to improve the performance of the model.

```
[In]: classifier = MultilayerPerceptronClassifier(labelCol=
'label', featuresCol='features', maxIter=100, layers=layers,
blockSize=128, seed=1234)
```

Now that we have defined every stage, we add all these steps to the pipeline and run it on the training data.

```
[In]: pipeline = Pipeline(stages=indexers + [featuresCreator,
classifier])
 [In]: model = pipeline.fit(train)
```

We now calculate the predictions of the model on train, test, and validation datasets.

```
[In]: train_output_df = model.transform(train)
[In]: validation_output_df = model.transform(validation)
[In]: test_output_df = model.transform(test)
```

```
[In]: train_predictionAndLabels = train_output_df.select
("prediction", "label")
[In]: validation_predictionAndLabels = validation_output_
df.select("prediction", "label")
[In]: test_predictionAndLabels = test_output_df.select
("prediction", "label")
```

We define three different metrics, to evaluate the performance of the model.

```
metrics = ['weightedPrecision', 'weightedRecall', 'accuracy']
 [In]: for metric in metrics:
    evaluator = MulticlassClassificationEvaluator
    (metricName=metric)
    print('Train ' + metric + ' = ' + str(evaluator.
    evaluate(train_predictionAndLabels)))
    print('Validation ' + metric + ' = ' + str(evaluator.
    evaluate(validation_predictionAndLabels)))
    print('Test ' + metric + ' = ' + str(evaluator.
    evaluate(test_predictionAndLabels)))
```

As you can see, the deep learning model is doing reasonably well on the test data, based on the input signal.

[Out]:

```
Train weightedPrecision = 0.976101874447846
Validation weightedPrecision = 0.9765821626938243
Test weightedPrecision = 0.9747324280445043
Train weightedRecall = 0.9755751041220662
Validation weightedRecall = 0.9761613691931541
Test weightedRecall = 0.9742582305920606
Train accuracy = 0.975575104122066
Validation accuracy = 0.976161369193154
Test accuracy = 0.9742582305920607
```

Conclusion

This chapter covered the internals of the basic building blocks of neural networks—artificial neurons—and the entire training process of a neural network. Different ways in which deep learning models can be constructed were mentioned, and, using Spark, a multilayer perceptron model was built.

Index

© Pramod Singh 2019
P. Singh, *Learn PySpark*, https://doi.org/10.1007/978-1-4842-4961-1

Printed in the United States
By Bookmasters